Mineral Resources and the Destinies of Nations

Walter Youngquist

National Book Company
Portland, Oregon

ISBN 0-89420-268-5

341330

Library of Congress Catalog Card Number 90-60670

Printed in the United States of America

Dedicated to geologists and engineers
the world over,
who find and produce the mineral base for our
modern civilization

—and—

to my wife, Elizabeth, and children
John, Karen, Louise, and Bobby

Acknowledgments

The able, very time-consuming (but willingly given) critical review work by Mr. William Eaton, Dr. Laurence Kittleman, and my former colleague at the University of Oregon, Professor E. M. Baldwin, during the progress of this manuscript has been exceedingly helpful and is deeply appreciated.

The numerous studies of regional and world petroleum supplies by Dr. Charles D. Masters and his associates at the U.S. Geological Survey have also been most useful as source material. Professors Peter J. Reilly (Chemical Engineering), and Garron O. Benson (Agronomy) of Iowa State University promptly, at my request provided basic information on certain aspects of energy economics as they pertain to agriculture. Professor William Purdom of Southern Oregon State College contributed information on radon gas and its geological associations.

For some global political concepts of mineral resources, I am indebted to my long-time friend Dr. Arthur A. Meyerhoff who, through both his publications and conversation, in large measure impelled me to pursue this present study. Additional and vital world-wide petroleum statistics have been generously provided through personal conversation, correspondence, and publications by Joseph R. Riva, Jr., Earth Science Specialist for the Science Policy Research Division of the Congressional Research Service, Library of Congress. His studies, made available to me, have been extraordinarily useful. A careful and thoughtful review of the entire manuscript by David S. Brown, Associate Director of the U.S. Bureau of Mines, and use of some of his public address materials made available to me, added valuable technical expertise and review to this study. Dr. Thomas Stitzel, Dean of the School of Business, Boise State University, read the final manuscript and offered his perspectives as an economist. Liston Hills, retired Chairman of the Board of ARAMCO, examined the sections dealing with petroleum, the Persian Gulf area, and particularly Saudi Arabia; and made some most important corrections and comments.

I also express my gratitude to several of my former employers who, from my studies for them, have given me the opportunity to acquire a view of both domestic and international mineral resource economics, as well as the political problems of mineral development.

Acknowledgments

These include the U.S. Geological Survey, and my consulting clients, the Sun Oil Company, the Shell Oil Company, Pan American Petroleum Corporation (Amoco), Belco Petroleum, and my former full-time employer overseas, International Petroleum Company, Ltd. of Peru, and later my consulting client, the Minerals Department of Humble oil and Refining Company — these latter two companies being divisions of what is now Exxon Corporation.

A consulting relationship now of more than 15 years' duration on geothermal energy for the Eugene Water & Electric Board has given me the opportunity to gain an insight on both national and international aspects of that interesting energy source.

Without these associations and the perspectives gained therefrom, I could never have written this book.

Finally, I am deeply indebted and very grateful for the initial acceptance of the manuscript and the subsequent encouragement through all phases of preparation of this book by Carl Salser, Editor, and Mark Salser, Associate Editor of the National Book Company. I could not have worked with more capable, considerate, and helpful people.

Eugene, Oregon

April, 1990

Foreword

From whatever time one wishes to recognize the beginning of human existence, to only a few thousand years ago, the demands of people upon the energy and mineral resources of the Earth were modest, almost negligible. But gradually, and then with a great rush at the time of the Industrial Revolution, energy and mineral resources became and remain the foundation for world development. This has been accompanied by a tremendous growth in population, which continues to the present time. The result has been a demand for these basic resources without historical precedent and a corresponding, world-wide endeavor by the industrial nations to obtain sources of these materials.

There has also been a rapidly increasing urgency to obtain foreign raw materials by the industrial nations. How rapidly a supply/demand situation can change is well illustrated by the United States in the case of crude oil. In 1960, the United States was a substantial net exporter of oil. However, by 1970, the supply of crude oil was just equal to demand; and by 1977, the United States was importing more crude oil than it produced. The oil crisis of the 1979-80 briefly raised prices and reduced demand for oil, so the United States (for a short time) produced more oil than it imported; but by 1989, crude oil imports again exceeded domestic production, a situation which is now likely to continue for the indefinite future.

The result of this, for a number of underdeveloped countries which had the good fortune to possess some of these resources in large quantities, has been to launch these nations abruptly from a backward location into the front lines of the twentieth century. An outstanding example of this is Saudi Arabia. A number of other countries also now have their economies dependent, to a great extent, on the exploitation of these mineral and energy mineral resources. This in turn has allowed these nations to develop some modern industrial complexes and also to initiate large social welfare programs (along with a considerable amount of corruption, in some instances).

But as production of these non-renewable mineral resources diminishes — and this is already beginning to occur in some areas — the question arises as to what will sustain the expanded industrial and social programs, together with the expanding population, when the mineral resources are gone? The story of the development and then the exhaustion of mineral deposits has already been played out many times in western United States and other mining regions, where ghost towns remain as monuments to a once affluent era. And with petroleum production declining in parts of the United States, once flourishing oil towns and even some states, such as Louisiana, are having economically hard times, because of the gradual loss of their energy mineral economic base.

The question might be as to whether or not this story will be played out on a larger scale, in some countries now very largely dependent on depleting mineral resources? Some of these situations are examined in this volume.

In addition to the demands on mineral and energy mineral resources caused by increased population, the highly specialized needs of new technologies have also put an emphasis on many minerals that previously were not generally used, some of which are relatively rare. Computers make use of beryllium, gallium, germanium, lithium, the platinum group metals, quartz crystals, rare earth minerals, rhenium, selenium, silicon, strontium, tantalum, and yttrium.

The modern jet airplane, besides demanding high quality aluminum for the body and wings, makes use in its jet engines of chromium, titanium, cobalt, manganese, nickel, and tantalum. Unfortunately for the United States, almost all of these vital jet engine materials are largely lacking, domestically. Several of them — such as cobalt, the platinum group metals, and chromium — exist in quantity, as commercial deposits, in a relatively few places which, because of various political considerations such as those, for example relating to South Africa and adjacent regions, create special international trade problems.

In the area of our energy industries (particularly in petroleum and in nuclear power), some 29 different non-fuel minerals are used. Telephone systems use 42 different minerals, and the electrical industry overall uses 85 different elements. In medicine, surgical instruments, chemotherapy, radiation treatments, and the many different diagnostic tests require a great variety of minerals. Somehow and somewhere these materials have to be found in the Earth, taken from the Earth, and by many complex procedures made into useful products for the benefit and welfare of the general public. Obtaining these materials from dependable sources and at reasonable cost is of great importance, particularly to the economies of the industrial nations, the leaders in advancing technology for the benefit of all humanity.

This volume looks at the impact of the present huge demand for minerals in both the industrialized and third world countries. The background information for this undertaking is broad, reaching into the domains of national and international economics, politics, technology, demography, history, engineering, and geology. Material for this study has been accumulated during more than 40 years, and in travels to some 70 countries. Yet nearly every day (and that is no exaggeration) there is another event which bears on these matters. Thus, it was necessary to draw an arbitrary line and put things down as of a specific date. This, then, is a snapshot of the race for survival and the pursuit of various degrees of affluence by the countries of the world — as of this moment. We also consider what technologies and mineral resources may be available for the future, and how these may affect various nations.

Besides being a record of minerals in world events to date, this is an attempt to make people aware of the past and present importance of these resources. Much of written history tends to ignore these fundamentals, perhaps because past historians generally were not educated in the realm of natural resources but rather were oriented toward social and political matters and did not recognize the very basic importance of the energy and mineral foundations of society. However, it is quite likely that historians of the present and the future will see this more clearly, as the energy and mineral demands of the industrialized world soar, and conflicts over them result — the Persian Gulf oil problems of the 1980's being an example.

The exponential growth of population and the exponential growth of the demand for energy and mineral resources during the last 100 years are now bringing this situation more and more into view. The two oil crises in the 1970's, for the first time, brought home to the general public in the United States the importance of oil, and what happens if it is not available. There was no real shortage of oil in the world at that time; the crises were politically motivated; but they did serve a very useful purpose in awakening people to the importance of energy mineral resources.

In addition to documenting the historical importance of mineral resources and making that apparent, at least to whomever may read this book, this study also presumes to look at the present scene and the future, with some degree of realism.

There is, I believe, a great need for this. To date, it has been possible with increasing knowledge, combined with yet untapped energy and mineral resources, to continue moving ahead with no great difficulty to attain the affluence which is enjoyed by some segments of world population. Because of this success thus far, when the question, for example, is put to the so-called "man in the street" as to what we are going to use for a convenient energy source once oil is unavailable, the common response has been "the scientists will think of something." We all hope they will, but at the risk of being thought a pessimist rather than a realist, one might observe that to date there were relatively simple steps going from wood to coal to oil, and then a slightly more complicated step in going to nuclear fission. But after oil, it appears we must either go back to coal (which has numerous drawbacks and would be a much more expensive source of liquid fuel than is oil) or we push nuclear fission. If either or both of those alternatives is not used, then presumably the move would be to fusion.

But the step from fission to fusion may be another matter. Surely it will not be accomplished very soon, on any substantial commercial scale. One estimate, perhaps overly pessimistic, is that it may never be accomplished, for it involves containing heat comparable to the Sun's core. There is the faint possibility that this estimate could be right. The announcement in 1989, by two electrochemists working at the University of Utah, that they had accomplished "cold" fusion, so far seems not to be verified to the satisfaction of the scientific community. Even if it does prove to be true, the time required for widespread industrial and citizen use of this technology would be measured in decades.

From the hope for ultimate use of fusion, as well as other energy sources comes a school of thought that says "there are more than sufficient renewable energy supplies to keep the world going indefinitely." (This is a "composite" quote of many such cheerfully optimistic views.) The energy sources for this utopia are cited as the nuclear fusion process, tidal, geothermal, hydrologic, biomass, and solar energy.

Here we presume to inject a note of realism. Hydrologic energy means dams, and the reservoirs behind them eventually fill up with mud. The number of good tidal sites is limited, even at best. Geothermal energy is quite site specific, and in a practical sense (for use in electric power generation) is a depletable resource, and even now is becoming so in places using it for direct space heating. Use of biomass, which is chiefly wood, is already resulting in deforestation of great areas, with consequent severe erosion and flood problems.

Nuclear power in the form of fusion (if such is ever commercially accomplished) is an almost infinite source of energy, but fusion may or may not become a reality. Solar energy is a low grade source which involves large capital costs if it is to be used on a significant scale. And if it is to be converted to electricity and widely used, as, for example, in transportation, it would also mean a tremendous change in lifestyle, as well as involving very formidable technological challenges.

The easy technologies have been done and the easily exploitable mineral resources are now rapidly being used up. We are coming to the end of one era and about to enter another.

This book looks back in history to view how minerals and energy minerals have played an increasingly significant role in human progress; and we examine their great importance today. As we look to the future, we indeed hope that the "man in the street" is right and that the scientists and engineers "will think of something." The latter part of this book considers those prospects. The challenges, however, for the first time in history, are global; and they are much larger and more difficult than they have ever been before. The scientists and engineers surely have their work cut out for them. And they will need help and understanding from the politicians. Hopefully, this volume may provide some background for such understanding.

On a minor editorial note, the term "petroleum" technically, and as used here, includes both oil and gas; the term gas means natural gas, not gasoline. Also, oil and gas, coal, shale oil, and oil sands are all minerals, just as is iron ore; but they are sometimes split off from other minerals by terming them "energy minerals." This has been done frequently in the book; but for the sake of brevity at some places, the term "mineral" has been employed to mean all these materials, both energy minerals and other Earth-derived materials. Also, the term "mineral" can be used for both natural chemical compounds, such as hematite, which is iron and oxygen, and for the element, iron, itself. Thus elements, such as copper, lead, and silver are regarded as minerals, as well as the more complex ores in which these materials may be found.

Table Of Contents

Mineral Resources & the Destinies of Nations

Chapter 1

Minerals Move Civilization

The market for sand dunes is not great. One hundred years ago a loosely organized group of nomadic tribes, a few small farm areas, and some little fishing villages occupied the Arabian Peninsula. The economic impact of this relatively small number of isolated people was negligible in the world economy. There was no indication then that things would ever be any different. Less than one hundred years later, these people, still a very minor percentage of the world population, violently shook the world's greatest industrial nation, the United States. Now there is, and will continue to be for many years, a very considerable concern by the industrialized nations not only about the happenings on the Arabian Peninsula but in all other countries around the Persian (Arabian) Gulf.

Commitment of international naval forces to this area in the 1980's was based solely on the matter of oil supplies. Otherwise, France, Great Britain, Italy, the United States, and the Soviet Union would not have been there. If oil were not involved, the war between Iraq and Iran would have been of little concern to the rest of the world. But the presence of oil resources (amounting to more than 60 percent of total world reserves in the nations surrounding the Persian Gulf) makes that area of current and continued importance.

We are sometimes told that we in the United States are entering a "service economy," but it seems doubtful that we can really survive as an important nation by serving hamburgers to one another, and exchanging computer printouts. The computers are made, at least partly, out of metal, and the energy to run them and fry hamburgers must come from somewhere. Hamburgers and computer printouts do not a great nation make. Energy resources and minerals are basic to any industrialized society. Who has had these resources available in the past, to win wars and to build the present state of civilization, has been vitally important; and who has these resources (or free access to them) in the future will be just as important. There was a recent brief example of this in the oil crises of the 1970's, in the United States.

Human history is sometimes even categorized in terms of rock and mineral resource-use stages — the Stone Age, the Copper Age, the Bronze Age, the Iron Age, and now the Atomic Age, although the correct term for this present time should be the Uranium Age, for it is that metal which produces atomic energy. It may be easier to simply recognize two ages — the Stone Age, and the Age of Metals. Africa, for example, went from the Stone Age to the Iron Age without the intervening steps. But in any event, use of minerals has strongly influenced the march of civilization from the Stone Age to the present.

The importance of mineral and energy mineral resources cannot be over-estimated. Most critical among the resources is energy, for energy is the basic physical wealth of the material world. Energy is the key which unlocks all other natural resources. Without it, the wheels of industry do not turn, no metals are mined and smelted, no cars, trucks, trains, ships or airplanes could be built, and if built, without energy they would not move. Houses would remain cold and unlighted, food would be uncooked, fields would be neither plowed nor planted; and the military defenses of the countries as we know them today would not exist. Without energy resources we would literally be back in the Stone Age. And without the use of energy and metals, as we employ them today, it is probable that the world's population would be reduced at least one-half, some estimates say 90 percent.

We in the industrialized world take our high standard of living very much for granted, without stopping to realize what a great debt we owe — almost every moment of everyday — to mineral and energy resources. We convert minerals to metals and, in turn, the metals into tools and machines. We run these machines with energy. By means of these we can produce easily and in great quantities the goods which in times past either did not even exist or took very large amounts of hand labor in their production. Minerals and energy in the form of a steel plow and the tractor and fuel to move these have allowed less than six percent of the working population to feed all of the United States and millions of others, too.

In contrast, 80 percent of the people of China still till the fields. Tilling the fields is important but when 80 percent of the population has to do it, it means that existence is close to simply the subsistence economic level for most of the population. The activity of tilling the fields has gone on for centuries; but only when the burden and necessity of making a living from the soil by dawn-to-dusk labor is at least partially lifted can human societies develop significant numbers of scientists, engineers, doctors, and technicians and thereby move society forward in a significant way. Tilling the fields precludes going to school, which provides the foundation for nearly all progress. Unnecessary manual labor in the fields and factories limits the human potential. Minerals and energy have allowed those portions of the world which have been able to obtain and use these resources to ease the physical burdens of existence and provide the time and the opportunity for the human mind to become educated, and that is where progress must begin.

In the factories and research laboratories, and in the service industries, too, energy continues to be all important. Employment and energy consumption, when graphed, are essentially parallel lines. This relationship between energy supplies and employment was strikingly demonstrated during the 1973-74 Arab oil embargo against the United States. As a result of that partial cutoff of oil to the United States, the Federal Energy Administration estimated that the nation's gross national product declined $20 billion, and a half million American workers lost their jobs.

The "energy mineral," oil, also allows products of our farms and factories to be widely and inexpensively distributed. It would do no good to produce quantities of appliances, clothing, machinery, or farm products, and all other things which form the basis for our material standard of living, if these could not be easily and cheaply transported to the population at large. This is what one of the energy minerals, oil, now does for us.

For many years the United States has stood at the head of the list of nations in terms of per capita consumption of energy and mineral resources. The United States, with about six percent of the world's population, uses about a third of the globe's annual energy supplies. Currently, in terms of energy, each U.S. citizen uses the equivalent of 300 slaves. In ancient days, the capture and use of slaves was one of the chief sources of wealth, and this continued even in the United States until the Civil War. Now we capture slaves in the form of barrels of oil from an oil well, cubic feet of gas from a gas well, tons of coal, or pounds of uranium ore from a mine. These, in a very real sense, are our modern slaves.

In regard to mineral and energy mineral resources, the U.S. Bureau of Mines has calculated that each U.S. citizen annually accounts (in consumption terms) for about 1300 pounds of steel and iron, 65 pounds of aluminum, 25 pounds of copper, 15 pounds of manganese, 15 pounds of lead, 15 pounds of zinc, and 35 pounds of other metals, such as cobalt, a material, incidentally, without which a jet airplane could not fly. The energy to go along with this use of metals includes 8000 pounds of oil, 4700 pounds of natural gas, 5150 pounds of coal, and 1/10th of a pound of uranium.

Including sand, gravel, cement, dimension stone, clay, and the energy and metal supplies just listed, more than four billion tons of new minerals are needed every year in the U.S. economy. We emphasize "new mineral supplies" because we continue to deplete the Earth's resources. These demands add up to more than 20 tons of raw energy mineral and mineral supplies which have to be produced each year for every man, woman, and child in the United States. These figures, of course, do not mean that each person directly uses this quantity of materials; but dividing the population into the total amount of these supplies used in the United States, annually, provides the figure. The steel, for example, may go into building construction (such as supermarkets or shopping malls), roads, bridges, cars, and trucks, all of which in one way or another serve citizens every day.

This total, then, represents our current material standard of living. As we add population, we have two choices in terms of how we live: either we dig up and produce more of these minerals, we import more each year, or by a combination of the two, we obtain that total of 20 tons of these materials for each citizen — natural born, legal, or illegal alien — or we reduce our material standard of living. The problem is that our population continues to increase, but resources are finite. Another important fact is that in the United States, in particular, and also in the world, in general, we use up our most easily won, higher grade mineral and energy mineral deposits first. Thus we are faced with an increased demand each year, against resources which are declining in quality and cost more to obtain.

Again, it should be emphasized that these are annual rates of consumption. Each 12 months, another 20 tons per person must be obtained, to simply maintain the status quo. The task is huge and getting larger each year as the population increases.

The desire for an increased standard of living, coupled with the continuing rise in population, have combined to put an exponential demand on mineral resources. In the first fifty years of the twentieth century, the total production of minerals and mineral fuels in the world was far greater than the total of such materials produced during all previous history. Then in the next twenty years following, this production was exceeded again by approximately 50 percent. These statistics, when graphed, show an exponential curve which is now beginning to steepen into a vertical line. Such a rate of consumption, of course, cannot be sustained. This strongly rising line, representing both energy mineral and mineral use, also represents the marked rise in the standard of living for the world in general, and especially for the industrialized nations. But to produce this steep curve of mineral and energy mineral resource consumption, the domestic resources of these materials in the industrialized nations have been drawn upon very heavily. The result is that many of the remaining and generally higher quality reserves of some of these materials are now held by the so-called under-developed nations, which are still in, or have just recently come from an agrarian or nomadic economy. Saudi Arabia is an example of this, as is Libya, and the United Arab Emirates.

The standard of living, as is generally defined, can be quite well measured by noting the consumption of two basic resources — iron and oil. Oil is the largest single energy source of the industrial or so-called first-world nations. Iron is the chief metal used. The consumption of oil is taken up in the chapter *The Petroleum Interval.* In the case of iron, at present, the 18 developed nations, with a total of about 700 million people, use iron (and its derivative, steel) at rates which range from about 680 pounds to about 1400 pounds per person per year. This compares with some 1.8 billion people, in a portion of the undeveloped world, who use less than 55 pounds of steel per person per year.

In the following chapter, we look at the role of mineral and energy mineral resources in history. The gist of the study is that these materials have been the cornerstone on which civilization has been built. Mineral resources have also been the source of wars, and during peacetime have caused great migrations of people, opening up new frontier areas — witness the gold rushes to California, Alaska, South Africa, and Australia. The rise and fall of both small and large communities have mirrored the discovery and development — and then the ultimate depletion — of a mineral resource. Mining ghost towns abound. Pursuit of mineral resources was one of the prime moving forces in the initial invasion by Europeans (chiefly the Spanish, in search of gold and silver) of both North and South America. The subsequent influx of less plundering peoples was also induced by the existence of the basic resource of land. But very important was the fact that the development of these raw lands was greatly expedited by the presence of mineral and energy mineral resources in North America — in particular, in a variety and abundance virtually unmatched by any other area of the world.

With the aid of this marvelous spectrum of mineral and energy resources, the United States rose from a wilderness to become the richest and most powerful nation in the world, in less than 200 years — an event unequaled in all history, and probably never to be repeated anywhere again. Just prior to this, and, in part, overlapping the rise of the United States, Great Britain, with moderate quantities of coal, iron, tin, and lead, in relatively close and convenient geographic association, had led the way into the Industrial Revolution. But Great Britain began to deplete these

17

modest quantities of mineral supplies in a comparatively short time. The much greater abundance of these resources in the United States, and also the generally higher quality (especially in terms of coal, and iron, and also ultimately in the discovery of vast oil deposits) allowed the United States to quickly overtake Britain and then surpass it in industrial development. Possession of energy mineral and mineral resources by the United States was the key to this phenomenal event, which deserves, and has been taken up in, a separate chapter, *The Luck of the USA and Saudi Arabia.*

But the United States, also, as in the earlier case of Great Britain, achieved a large and fast rise in its standard of living at the expense of the rapid exploitation and depletion of its mineral and energy mineral resources. The United States is the most thoroughly drilled country in the world, in terms of oil and gas wells. Onshore, and to a considerable extent offshore in friendly waters (the Gulf of Mexico and offshore California, as contrasted with offshore Alaska), we are running out of places to drill.

Onshore in the United States there is hardly space between dry holes or already producing or depleted oil wells for the discovery of a major oil field. By major, we mean a field of 100 million barrels or more (and it should be noted that a 100 million barrel oil field, a major find, indeed, would be less than six days of current U.S. oil consumption — we simply do not have, nor can we find a 100 million barrel oil field every six days)! We are now, for the most part, picking away at very modest-sized oil reservoirs which are left in the hospitable areas of the 48 adjacent states.

Beyond that, we are forced to go to much more difficult and expensive frontier areas, such as the deeper waters of the Gulf of Mexico and the sub-arctic and arctic regions of Alaska, to find places yet undrilled and which are large enough to be the potential site of a major oil discovery.

In marked contrast, the nations of the Persian Gulf region have no climatic and logistic problems to deal with in developing their oil reserves. The weather is warm and it is easy to bring in supplies and equipment. And no long pipelines over hostile terrain (such as the 900 mile Alaska line) are needed to carry out the oil. It is also noteworthy that the average production per well per day in the United States is about 14 barrels. This is quite different from the more than 12,000 barrels a day per well being produced at the Ghawar Field in Saudi Arabia.

In 1970, a most significant event took place in the United States, largely unnoticed by the general public. In that year, the

18

curves of oil consumption and oil production crossed at about 11.3 million barrels a day (This includes liquids condensed from natural gas.). Prior to 1970, the United States had been self-sufficient in oil, and, in fact, for many years, a large exporter of oil. In 1954, for example, when the Egyptians closed the Suez Canal, disrupting Britain's oil supplies from the Middle East, the United States, with surplus production shut in, simply opened its oil well valves a bit wider and took care of the problem.

Going back to 1909, the United States that year produced 500,000 barrels of oil a day. This was more than the rest of the world (combined) produced at that time! Remarkably, the United States continued to produce at least one-half of the world's oil until the early 1950s. At the present time, however, the United States produces about 9 million barrels of oil and gas condensate a day. This is now less than 20 percent of world production, and this percentage can only go down — which it is. In less than 50 years, the United States went from being the world's largest oil exporter to being the world's largest oil importer, the bill for which is currently nearly $50 billion annually.

As a nation, the United States tends to have a very short memory. The long lines of cars at gasoline stations, in the 1970's, were speedily forgotten in the temporary glut of oil in the middle and late 1980's. But oil is a finite resource, facing a potentially infinite demand; and whoever owns the oil reserves in the future will have a lot to say about the future of world economies. Unfortunately for the United States, its reserves have been steadily declining for a number of years. Reversing this trend is probably not possible, at least for any length of time. Actually, in 1987, there was an apparent reversal of this trend, as total oil reserves were raised. However, this was done in part by drilling around the edges of known fields rather than through substantial new discoveries, and partly by revising the reserve figures of already discovered fields. This was done on the premise that more oil was being recovered than had been previously estimated, but there is some question as to how much of this book-keeping entry will prove to be justified. In any event, it is essentially a one-time change and oil reserves cannot be increased every year by simply adjusting the figures.

After the milestone year of 1970, the United States has had to import an increasing amount of oil; and it can be flatly stated, without any fear of contradiction (a risky statement but we stand by it), that the United States will never again be self-sufficient in oil. That day is gone forever. The effect of this circumstance is and will continue to be very important in shaping the future of the United States, for it not only affects the vulnerability of its industrial economy but it also affects its international monetary position through the balance of trade accounts, and the national debt. Imported oil costs money, and the cost of this oil is already aggravating the balance of payments problem, which has become very large.

Regarding metals, the story is much the same. The two percent (or in some cases locally much richer) copper ores derived from the native copper deposits of the Upper Peninsula of Michigan are just a memory. Only one mine survives along with cold smelter smokestacks and rusting mine headframes, standing starkly in the woods. The present reality is that the ½ of one percent copper ore we now have available to mine has a hard time competing with the higher grade ores available in Peru, in the giant deposits of Chile, and in certain African nations. The same is true of the rich hematite iron deposits of the once mighty Mesabi Range of northeastern Minnesota. The 60 percent or better iron content hematite is gone and the taconite with only about 30 percent iron (which must be beneficiated) is what remains.

We have used the United States as an example of the depletion of mineral resources which has occurred in the industrial nations. But smaller nations have mineral resource depletion problems also; and these are just as important and in some cases more important to these nations as their mineral and energy mineral resources may be even larger in their total economy than in the case of the United States. Indeed some of these smaller nations are largely dependent on one or just a few mineral resources, and in some the end of these resources is clearly in sight. Such is the case of Nauru, an island nation in the South Pacific, which has large fertilizer deposits as its only resource, and these will be exhausted in less than a decade. What then? We discuss this later.

The abandoned copper mines and towns of the Keweenaw Peninsula of Michigan and hundreds of ghost towns of the American west are local situations, where the mineral economic base has been exhausted. As larger mineral and energy bases are eroded also, the same decline will happen as in the smaller communities, unless states or nations can develop other industries, establish a renewable resource base of some sort, or develop some special skill amongst its citizens which is in demand in a larger sphere.

Some of these larger areas even now are having their problems. Texas and Louisiana are having oil withdrawal pains, which will continue for a long time to come, as both states have peaked in oil production. How much of this can be countered by marshaling the ingenuity of the citizenry remains to be seen. When you drill a well, set pipe, and just open the valve and a valuable product flows out for years, that is easily-won wealth. When the option to continue doing that is no longer available, it is a vastly different economic situation, and so it is becoming for many areas.

However, for all we have just now stated regarding the importance of mineral resources, some nations which have no substantial mineral or energy mineral bases have been able to do very well. Japan, with very few energy or mineral resources, has become the economic miracle of the latter half of the twentieth century. Can that miracle be sustained or will Japan be the first major industrial nation to demonstrate the fatal flaw in having no indigenous sources of these vital materials? It was the lack of these energy and mineral resources and the cutting off of foreign supplies of oil, in particular, which caused Japan to go to war in 1941. Will Japan ultimately lose the industrial "war" because of a lack of raw materials?

Energy and mineral resources are vital to virtually all phases of an economy, be it the time-honored tasks of farming, construction, or transportation; or the newer high-technology areas of communications, where satellites and vast networks of electric grid systems stretch across nations and around the world. All this is made out of minerals of various sorts and uses vast amounts of energy. The initial process of xerography, which revolutionized the world of printing, was made possible only by the existence of the metal, selenium. Tungsten is critical in the high-speed, hard-tool steel used for cutting other steel — the cylinder blocks of automobiles are bored with tungsten carbide steel, and tungsten is the usual filament in light bulbs.

Whether it is air conditioning, which makes life livable and productive in hot, humid climates, or heat, which keeps us comfortable in cold weather and in cold climates, energy is vital. Many parts of the world would not be habitable to any large populations if it were not for energy supplies in quantity.

In national defense, everything from tanks, to submarines, supersonic planes, rockets and intercontinental ballistic missiles require minerals for their production, and energy to power them. In fact, in modern warfare it has virtually become a matter of which side can deliver the greatest amount of energy in some destructive fashion against the other side, which would determine who wins. A demonstration of this was made in Japan twice in August of 1945. We must hope that such a use of the energy mineral, uranium, will never occur again.

It is important to note that in the production of energy and mineral resources, other energy and mineral resources have to be used. Imagine drilling a 10,000 foot deep oil well by hand or breaking up iron ore with a pick and shovel. To drill wells; blast out ore deposits, load, and transport them; to manufacture drilling pipe and drill bits; to make oil well casing; to build turbines and generators for power plants; to make solar collectors; and to make and run the chain saw which is used simply for cutting firewood, energy and minerals have to be used.

It is also important to recognize that as we have to mine or drill deeper to obtain energy and mineral resources, or with the same expenditure of energy and minerals recover lower and lower grade resources, eventually there comes a time when the energy or the value of the material recovered is less than the cost in energy or in the material used to recover the resource. Until that time, the game can continue, albeit with less and less efficiency; but when the expenditure is greater than what is recovered, the game is up. Technological advances in recovery methods can delay that day but the trend is inevitable. There are already several studies of when the net energy loss for the operation will occur for oil and gas in the United States. The date differs from study to study, but all studies do come up with a date, and all the dates are before the year 2010.

Minerals, including energy minerals, are the basis for our modern civilization. Nations not possessing these are either doomed to stay at a relatively low standard of living, or they have to get these resources in raw or finished form by trade, in some fashion. If denied, they will have to resign themselves to a second class position or go "to war." This was precisely the decision which Japan faced in 1941. Can free trade and access to raw materials be maintained from now on, so that the "Japan decision" will not ever be repeated?

Competition for these vital resources exists today. This is the "cold war" or, as one former United States president chose to call it, the "real war." It involves both minor military moves (carefully designed, so as not to stir up any major struggle which might spread), and political moves directed toward ultimate control of the large quantities of crucial materials possessed, for example, by South Africa, and some of the mineral-rich neighboring countries, such as Zambia (cobalt and copper), Zimbabwe (chrome), Zaire (cobalt and copper), and Namibia (large uranium deposits, only partly developed). This situation, for the moment at least, seems to be generally ignored by the United States. But it has not been ignored by the Soviet Union, which continues to stir in these areas.

Soviet President Leonid I. Brezhnev confided to Somalian President Siad Barre, when that nation was for a time an ally of the Soviet Union: "Our aim is to gain control of the two great treasure houses on which the West depends — the energy treasure house of the Persian Gulf, and the mineral treasure house of central and southern Africa." And this thought was not original with Brezhnev, for the father of communism, Lenin, in his tract "Imperialism," published in 1916, described how the overthrow of capitalistic democracies could be accomplished by depriving them of their sources of energy and raw materials essential to their survival. The Soviet Union clearly sees the importance of the mineral and energy mineral economic bases, to the western world. The western world, however, long accustomed to the high standard of living which past exploitation of largely domestic resources has brought, seems generally unmindful that the local supply house is running short.

This book recalls the importance of mineral resources in history. It describes their importance today, tells who has them, and suggests what may lie ahead tomorrow for regions and nations as energy and mineral resources are depleted in one area, and how these resource "sources" shift geographically, and economies wax and wane accordingly. The question of how far technology and free trade might compensate for the unequal distribution of mineral resources is considered. As the mineral and energy mineral future unfolds, virtually every human being will be personally affected, including you. With the price of oil right now, whatever it may be, you are being affected.

Of special significance, worth emphasizing again, is the fact that as energy and mineral resources are depleted and costs go up, sources of these materials change, witness the shift of major world oil production from the United States to the Middle East. With these geographic shifts of resource centers also go balances of economic power among regions and nations. Hibbing, Minnesota was once the iron ore capital of the world; but the high grade ore was mined out. Much of the iron used by the United States to make the weapons to defeat Japan came from the Hibbing mining area. Japan currently gets much of its iron ore from excellent deposits in relatively nearby Australia. And residents of Hibbing buy Japanese cars made from Australian iron. How things change!

Such shifts in mineral sources have occurred in the past and will continue to occur in the future. The United States is gradually experiencing this truth, the ultimate outcome of which will take many years to play out; and the process will have a continuing and substantial effect upon the lives of all U.S. citizens. Yet many persons do not seem to realize that in the United States, the era of high grade energy mineral and mineral resource abundance is gone. Beneath all of history has been the need of every nation to obtain the basic necessities of life and endeavor to upgrade its standard of living. This has been the chief, underlying, moving theme in the story of civilization. It will continue to be so.

In World War I, it was said that the Allies "floated to victory on a sea of oil." In World War II, a previously obscure metal, uranium, in the form of the atomic bomb, dramatically ended that struggle. Currently, the USSR has the world's largest natural gas reserves and is extending pipelines to carry this energy into western Europe. But the valves which control these gas supplies are on the Soviet side of the border. Perhaps in the future, atomic war will not be feasible as all the world would be the loser. Ultimately, wars may become strictly economic. These wars, then, are to be won, not on the battlefield, but in the mines, and oil and gas wells of the various nations, or won by those nations which can gain access to vital energy and mineral resources held by other countries. Unfortunately, the United States appears not to be gaining in this struggle, for the USA, chiefly through the Department of the Interior and its spokesmen, states there is an energy and minerals policy. There may be such on paper; but in actual practice, the USA does not seem to be implementing any energy or mineral policy. In sharp contrast to this situation, the Soviet Union for many years has had long-term energy and mineral self-sufficiency as a major goal. And indeed they have done very well in this regard, for they are now the most energy and mineral self-sufficient nation in the world.

The fact of this near self-sufficiency of the Soviet Union, with regard to minerals and energy minerals (huge coal and natural gas reserves), compared with other industrial nations will be something to carefully watch. The Soviet Union is not now a world economic power, but if the current attempted reforms of their economic system succeed, they have the resource base to progress very well and very far. It should be noted, however, that the USSR is just now peaking in oil production; and self-sufficiency in that regard may disappear shortly. The USSR is the closest major nation to the oil-rich Persian Gulf. More military and economic maneuvering in that region seems inevitable.

Mineral resources have been, are now, and will be, for the indefinite future, an over-riding force in the future of nations. Let us proceed with the story.

Chapter 2

Minerals in History

A rock thrown at game, or bounced off the skull of an enemy was probably the first use of minerals by humans. Eventually the notion occurred to someone to chip a rock into something which could be used as a knife or scraper; and flint, because it can be easily flaked into such tools, became important. Flint occurs geologically in a variety of ways, but it is most readily found and extracted as nodules in limestone. In certain valleys of France, especially in the La Claise River Valley where, for several miles along the bordering limestone cliffs, flint nodules are present in great abundance, ancient peoples developed extensive flint workings, chiefly near the present village of Grand Pressigny. This probably was the most important mining district in Europe at that time.

The Stone Age lasted for a long time, during which flint was among the more valuable possessions a human could have. Then deposits of native copper were discovered, in both the New and Old worlds. Before that, however, gold was used as an ornament, for gold is commonly found native — that is, in pure metallic form, uncombined with other elements — and is therefore conspicuous. Bright and attractive, gold can be easily beaten into desired shapes and this was done at an early date in human history.

But the first "working metal" used by the human race was copper. Copper was extensively mined by early people in the Sinai Desert area and later in Cyprus. The deposits on Cyprus were so highly valued that war followed war in bloody contests for the metal; and the island passed under successive control of many groups, from the Egyptians on down through the Romans. The Romans gave us the word from which "copper" is derived, by shortening *aes cyprium* (Cyprium copper) to *cuprum.*

On the Upper Peninsula of Michigan (the Keeweenaw Peninsula portion) numerous prehistoric pits (in the extensive native copper deposits there) testify to the widespread use of this metal by American Indians. In fact, it appears that the mining and working of copper in this area was a very large enterprise. The presence of abundant native copper, which could be relatively easily worked into knives and other tools, was of great importance to the Indians, who otherwise had to use stones, which were much more difficult to work. The copper could be hammered into all kinds of implements, including arrowheads, chisels, hooks, and axes. More than 10,000 individual mines have been located in this area and it has been estimated that it took at least 1000 miners a minimum of 1000 years to produce all the copper workings now visible. This copper was apparently also a major basis for trade, as tools of native copper have been found in Indian habitations throughout the Upper Midwest, and down the length of the Mississippi River Valley.

Elsewhere in the world, native copper tools helped shape the stones of the pyramids of Egypt, and copper was the first metal employed as a shaped weapon in Old World warfare. Copper ores are relatively easy to smelt and so copper metallurgy came into being early and copper became the first metal to be used extensively by several civilizations. Its use marked a transition from the Stone Age into the age of metals in general.

Here we get ahead briefly in our story on copper to note that without copper the electrical age may never have come about or would at best have been long delayed, for copper has been the workhorse of the electrical industry, only recently being partially displaced by aluminum and glass fibers. However, the production of aluminum has been dependent on vast amounts of electricity produced by copper coil-wound generators and initially transported by copper wires to the aluminum smelters. Without copper, we might still be reading by candlelight or oil lamps.

Beyond the Copper Age came the Bronze Age. The metallurgist who discovered that tin, added to copper, would make it much harder is unknown; but this fact became known to the Romans who eventually took possession of the great tin deposits of Cornwall, in England. Roman weapons, made first of bronze and later of iron, conquered much of the then known western world. Although iron was known much before Roman times, it had only limited use, as the metallurgy of iron is difficult, due to the high temperature required to melt it. Because of this, for a long time iron was not well known nor widely used. However, there is a record of iron being employed as far back as 1450 B.C.; and about 1385 B.C., the Hittites manufactured a substantial number of weapons from iron. With this superior weapon, they went on to subdue the Assyrians, and then drove into northern Syria and Palestine. There they fought the Egyptians, ultimately going on to establish the Hittite Empire as a major political and military power in western Asia. But the iron which had brought the Hittites to power was eventually used against them. The Hittites had jealously guarded their secret of iron metallurgy, because of the great military importance of this metal. But the secret got out and ultimately the enemies of the Hittites were equipped with the same metal, and the Hittite Empire was besieged and finally disintegrated.

The Iron Age, as we know it, however, came much later, in the sense that iron became of widespread use in industry and construction fairly recently — during the last one thousand years or so. Iron really came of age in the Industrial Revolution, led by Great Britain. That country had the good geological fortune of having large supplies of both coal and iron ore, very close to one another, so they could readily be brought together to produce the iron and steel needed to build the factories and the machines so necessary to Britain's Industrial Revolution. The fact of this happy coincidence can hardly be overemphasized.

Coal did indeed carry Great Britain into the Industrial Revolution; but when oil became the leading (and more versatile and desirable) energy source, the United States took over the economic lead in the world, because of its huge petroleum deposits, combined with its other rich mineral resources.

Although we are now said to be in the Atomic Age, we don't construct buildings or make cars of uranium, we use steel and aluminum. So in a very real sense, we are still in the Steel/Aluminum Age, and we will probably be in it for the indefinite future.

We have traced very briefly the discovery and use of some of what are called the industrial metals, but there is another tale involving the great impact of metals on human history. This is the record of how, in many different ways, the search for and discovery of the precious metals, chiefly gold and silver, have affected history.

As we have noted, gold was the first metal worked by humans, as it is bright and attractive in its native form — the form in which it commonly occurs. It can easily be worked into many shapes and it does not tarnish. Gold nuggets in stream beds attracted attention very early. This attraction of organisms for gold goes back probably before humans, as pack rats and some birds will pick up small gold nuggets and put them in their nests. So the attraction for gold is apparently a basic instinct, and continues so to this day. Silver is rarely found in native form and it tarnishes easily. However, many silver ores can be smelted readily, so silver, too, has a long history of being attractive to humans and put into use among earlier peoples, chiefly as ornaments and later as coinage.

Records of the invasion of lands caused by the search for gold and silver, and how these metals thus have affected history have already filled many volumes. Gold was one of the earliest reasons for conquest and exploration. What is perhaps the first map ever made is a papyrus map of the Rammessides, which shows a route to the Coptos gold mines along the eastern border of Egypt, fringing the Red Sea. Egyptians reaped great quantities of gold from this area.

The gold occurred as native gold in white quartz veins, and in placers (sand and gravel deposits) downstream in the valleys below. But even richer gold deposits were in the upstream areas of the Nile, outside of Egypt in the Nubian Desert. In fact, the name "Nubian" comes from the Egyptian word "nub," which means gold. The Nubian gold lured the Egyptians to exploration and conquest of that area, and the first gold was simply taken by plunder from the natives who already had it. Along with the gold, the Egyptians took a host of slaves. Eventually the Pharaohs sent miners and established regular gold mining camps in Nubia. This precedent of exploration for gold plunder and then later the establishment of permanent gold camps was a pattern to be repeated many times later in history.

Egypt thus became rich in gold but was poor in basic raw materials for life. But it had at that time the world's largest gold-filled treasury, which enabled the Pharaohs to develop Egypt as the most powerful nation in the Middle East. With their gold they could buy whatever they needed.

Silver came into use somewhat later than gold, as it rarely occurs in native form and tarnishes quickly. But it is easily recovered from lead ore complexes, which attracted early attention, because the chief ore of lead is galena, which has a bright, shiny, metallic luster, is very easily smelted, and readily releases the silver with which it is commonly associated. Silver, in fact, even today is produced chiefly as a byproduct of the smelting of other metal ores. Much silver came from Asia Minor, and the Babylonian traders made it the first widespread standard of exchange. The unit of exchange was called the shekel, a name which persists to this day as the name for Israel's unit of currency. At that time, one unit of silver was valued at approximately 40 units of lead, or 180 units of copper.

Silver was discovered in many areas of the ancient world, but in one area, in particular, it played an important part in affecting the course of western civilization. In the limestone hills near the town of Laurium and the village of Plaka, about 30 miles from Athens, huge deposits of silver were discovered. For a thousand years, Athens and Greek culture, in general, flourished because of the wealth taken from these mines. Each citizen of Athens was given yearly a share of this treasure, recovered at great effort and loss of life, by many thousands of slaves working in the mines.

But threatening this wealth and culture enjoyed by the Greeks were the Persians, who first moved against Greece in the year 492 B.C., under Mardonius, who had a substantial fleet and many men. However, this Persian fleet was largely wrecked by a storm, as it approached southern Greece; and the survivors went back home. The threat did not disappear, however, but only grew larger as the Persians reassembled and made an attack upon Attica in 490 B.C., with a fleet of 600 ships.

After this invasion of southern Greece, the Persians went on to ultimately land on the beach at the plain of Marathon, only a few miles from Athens. Here, 20,000 Persians were met by only 9,000 Athenians and 1,000 Plateans. But the Greeks, with great courage, staged a mass running attack against the Persians which demoralized them and they fled in confusion. The Persian fleet retrieved the survivors. Reportedly, some 6,400 Persians were killed, whereas the Greeks lost only 192 soldiers.

But even with this defeat, the Persian menace did not vanish. The Persians planned still another invasion and this is where silver played a crucial role. Thermistocles, a perceptive and foresighted Greek, persistently pushed for a substantial navy for the Greeks. To accomplish this, he suggested that the Athenians forego their annual dividend from the great silver mines near Athens; and that the money be used to build ships. At the time, the Greeks had only about 70 ships; but the Athenians heeded Themistocles and ultimately 130 more ships were built with the silver.

The Persians did come again with a great fleet and several hundred thousand men. After a series of skirmishes, the Persians took and sacked Athens. It now remained for the Persian fleet to destroy the Greek navy, which consisted of about 300 ships — 200 of them Athenian, 130 of which were built with the silver from the Laurium area — these were the newest and best ships.

In late September of 480 B.C., the Persian commander, Xerxes, took a seat on a prominence overlooking the Bay of Salamis, just west of Piraeus, the port to Athens, and confidently prepared to witness the destruction of the Greek navy.

The ships of the Greeks were smaller and outnumbered about three to one by the Persian vessels; but the Greek ships were very easy to maneuver, and they were fitted with battering rams. The Greek ships quickly moved into and through the Persian fleet, shearing off their oars, ramming them, and leaving them dead in the water, waiting to be further rammed and sunk. Figuratively and literally the Battle of Salamis marked the high water mark of the

Persians in Greece, and eventually the Greeks freed themselves from the Persians. Thus Greece continued to flourish and provide the foundation of western civilization in the form of its democratic ideals, and the corresponding liberties in individual freedom and in economics. Without the silver mines to provide the ships with which the Greeks destroyed the Persian fleet, the course of history would have been markedly different.

About 150 years later, Philip of Macedonia (a province in northern Greece) captured the Athenian town of Amhibolis which was key to an area rich in gold and silver. From this area, great quantities of silver and gold were taken by Philip, and with these precious metals, he hired and trained a large professional army. He used this army to move south and ultimately united much of Greece to fight against their ancient enemy, the Persians. In 336 B.C., however, Philip was murdered, and his son, later to be known as Alexander the Great, took over the leadership. Alexander continued to draw on the mines in Macedonia and Thrace, and with silver and gold from these sources, with which to pay his army, he began to move eastward against the Persians.

The Persians, in their conquests, had accumulated immense wealth. This Alexander wanted, and ultimately he did capture the Persian treasury. Plutarch reported that it took 20,000 mules and 5,000 camels to transport it. These precious metals were carried from the eastern world to the Greek western world and made into coinage, which financed a great expansion of trade and commerce, and, in turn, the arts. Immense quarries were opened in the marble hills of Greece. The resulting — and now world famous — sculptures, made from these rocks, record the golden age of Greece, largely financed, first, in effect, by the silver from Laurium and vicinity, which allowed the Greeks to defeat the Persian invasion. Then, with the additional Greek silver and gold which financed Philip and his son, Alexander, the Persian treasury was captured and brought to Greece, to further sustain Greek culture and Greece as a major force in the developing western world.

The ability of the Romans to ultimately develop an iron-based weaponry, and the depletion of Greek mineral resources and its treasury, finally led to the defeat of Greece and the transfer of what was left of the Greek treasury to Rome. But Rome also developed problems based on precious metals or lack thereof. Unable to take advantage of the present day method of government finance, because the printing press had not yet been invented, the possession of the metals, gold and silver, was all important. This is taken up in the chapter on minerals and money (Chapter 5) in more detail; but in brief, the Italian peninsula itself had very little in the

way of metal, and especially precious metal resources. The precic
metal mines of Spain were a major factor in supplying the gold and
silver which made the Roman currency of value. But as these mines
were depleted, the Romans began to debase their currency, eventually to the point where it became virtually worthless. The Romans
ultimately were unable either to pay their mercenary armies or to
import the goods needed to sustain their standard of living. There
were other reasons involved in the fall of Rome, but debasement of
its currency was an important contributing factor.

A minor power in the Mediterranean, the Venetians, made a
good thing of their only mineral resource, salt. With this mineral,
they were able to become fairly extensive traders and extend their
influence well beyond their tidelands home city. A substantial
Venetian fort exists at the entrance to the harbor at Iraklion, in
Crete.

One of the most striking and well known aspects of minerals
in history has been the way in which the search for and discovery
of these (chiefly gold and silver) has caused mass migrations of
people. In many cases, minerals precipitated the opening of new
lands. It has been said that "the flag follows the pick." The search
for gold and silver lured the Spaniards to the New World, resulting
in the conquests of Mexico, Peru, and some of the adjacent lands.

It also was said that the Spaniards had the "gold disease";
but the sad fact is that they had other diseases, too, such as
smallpox. Many of the native cultures the Spaniards encountered
were virtually destroyed by the diseases which the searchers for gold
and silver brought with them. In that unfortunate way, too, gold
and silver further altered the history of the Indian nations of the
western hemisphere.

The discovery of diamonds in South Africa, in 1867, brought
thousands of immigrants to that region. The subsequent discovery
of many other minerals of importance, most prominent among
which was gold, led to the transformation of what was largely an
agricultural economy in South Africa into the present day urban-
industrial economy. Concurrently, the continuing international
interest in South Africa, over the years, has stemmed largely from
its possession of several strategic minerals. It remains a potentially
valuable prize.

The discovery of gold in Australia likewise opened up large
areas of that continent, which previously had been ignored. The
early gold rushes to Victoria and New South Wales caused many
changes in Australia. In 1850, there were only about 400,000

people in all of Australia, but by 1861 there were more than a million. Melbourne got its start during the gold rush, and at one time was the richest colony in the British Empire. Many of the fine gardens and some of the buildings (such as Government House) are legacies from what the discovery of gold, in the mid 19th century, did for Australia.

The gold rush to California is one of the great epochs of migration caused by a mineral discovery. Ships went around Cape Horn; others went to Panama, where the passengers hiked across the Isthmus, to pick up another ship on the other side. Still others took the overland route to California. California was really opened up by the discovery of gold; and it went on to become one of the great states of the United States, and developed an economy of global significance. This would have happened in any event, but gold was the initial catalyst.

The Yukon and Alaskan gold rush of 1897-1898 was the last gold rush of the 19th century; but it had all the excitement and problems of the previous gold rushes, and it opened up virgin territory. It had its origin when two prospectors, Robert Henderson and George Carmack, were salmon fishing in a tributary of the Yukon River, the tributary later to be called the Klondike. These men saw the glint of gold in the stream bed late in the summer of 1896; but the news of the discovery did not get out until 1897, and then the rush was on.

Dawson City rose from almost nothing to a population of 25,000, within a single year. By February, 1898, 41 ships made regular runs between San Francisco and Skagway, Alaska, which was the port nearest the goldfields. From Skagway, the prospectors had to go over either Chilkoot Pass or White Pass, to the Yukon. During the winter of 1897-1898, 22,000 people were checked through the border between Canada and the United States, on these routes.

But there is another saga even more striking and important than these relatively localized migrations of peoples to mineral localities. It is the story of the incredible luck of the United States, wherein there can hardly be a more striking example of the importance of minerals to a nation. A marvelously rich and varied spectrum of mineral and energy mineral resources was progressively discovered in the United States. By this circumstance, the United States was able to rise from a raw wilderness, with a minute population, to a position of the most powerful and richest nation on Earth, supporting a population of 200 million people, in the span of not much more than 200 years. It is an event, however, which can never be repeated, for there is no other virgin continent which can be discovered and exploited; and indeed to accomplish this great and spectacular rise in wealth and importance, the United States has gone through its mineral resources with a speed also unparalleled in world history. This story merits part of a special discussion, which is Chapter 9 of this book: *The Luck of the USA and Saudi Arabia.*

Minerals have played a very important part in warfare, especially the more recent conflicts between industrialized societies. This large and critical topic is dealt with separately in the next chapter. Suffice it to note here that it is doubtful, for example, that Germany would have embarked on World War I or II or that Japan would have precipitated war with the United States on December 7, 1941 if these countries had, within their own borders, a diversified and ample mineral (including energy mineral) resource base. Clearly, in the case of Japan, which has very limited resources, the perceived need to go to war was based on gaining access to mineral and energy mineral resources. The immediate decision to go to war was precipitated by the cut-off of oil supplies only a short time before, by the United States, and by what was, at that time, the Dutch East Indies.

Finally, it should be noted that historians tend to write history in terms of political intrigues and alliances, treaties and diplomatic maneuverings; but these are relatively superficial matters which, on deeper analysis, can usually be seen to have involved a struggle over basic natural resources, upon which all human life must depend.

The role of minerals in history has thus been relatively obscured to date. But more recently, particularly due to the two world wars which have depleted many mineral deposits, the several oil crises which the western industrialized world has experienced, and the troubles in the 1980's in the Persian Gulf, the role of minerals, particularly energy minerals, has become much more visible. The long lines at the gas stations in the USA, twice during the 1970's, brought home the importance of energy mineral resources to nearly everyone.

As these resources, which are finite, are depleted but populations, which do not yet appear to be stabilizing worldwide, continue to rise and raise the demand against finite mineral resources, such resources will be of increasing national and international concern. Gradually it will become apparent to the general citizenry that the availability and use of minerals, and energy minerals in particular, very largely controls the way of life and the future of the human race.

For a brief time, at least, the citizens of the United States became aware of this during the 1970s gasoline shortages. But unfortunately that experience faded in memory with the oil surpluses of the 1980s. That respite, however, is only temporary. Access to minerals and to energy minerals determines the standard and mode of living today, although with relatively few shortages, this fact is not headline news at the moment. In times past, minerals have played a lesser role in the lives of citizens, in general, for in a small farm and cattle-grazing economy, which much of the world had for centuries, there was no great need for a widespread use of metals. Energy sources were there, it is true, but people did not know how to harness these servants. Engines had not yet been invented.

But when industrialized societies developed in the world, at that point minerals and especially energy minerals became essential to the lives of a great many people. If those of us who live in these industrialized societies try to maintain the affluence we now enjoy, minerals will remain central to our economies, many of them critically so. How we can continue to have these minerals and energy minerals in the quantities needed, and pay for them, are problems which will demand increasing attention in the decades to come. People are not likely to surrender the living standards to which they have become accustomed, without a struggle. How to prevent such struggles will be a challenge for scientists and engineers, economists, statesmen, and for society at large.

Chapter 3

Minerals and War, & Economic Warfare

Initially, raw manpower, in the form of hand-to-hand combat, with perhaps a few minerals in the form of rocks, were the ingredients of warfare among early humans. The stone axe and knife were invented; and then the longer knife, called the sword, and the spear were devised.

The first knife was stone but when native copper was discovered and hammered into knives, as was done in several places in the world, the metal age of warfare began. Metals in war then became important and they have been, ever since.

Arrows had a similar history, as the hundreds of thousands of stone arrowheads testify. Arrows were first tipped with materials such as flint and obsidian. It is known that Indians fought for the possession of Obsidian Cliff, in what is now Yellowstone National Park — probably one of the earliest battles fought for the possession of mineral resources in the western hemisphere. It also may have been that when obsidian was discovered and found to be razor-sharp when flaked, that the Indians concluded "now that we have the obsidian arrowhead, war is too terrible to contemplate," but civilization nevertheless progressed.

The wonderfully rich copper ores of Cyprus were the cause of numerous ancient conflicts. The Huns fought for the possession of salt deposits in southern Germany. The Phoenicians and later the Romans fought to control the tin mines around Cornwall, in England. France and Germany bitterly contested for the iron ores of the Alsace-Lorraine area, and battles were fought over the lead and zinc deposits of Poland. And as part of the settlement after World War II, Poland received some of the coal mines which previously were in German territory.

The desire by Chile to have part of the Peruvian/Bolivian nitrate deposits, in the Atacama Desert, was the cause of the Nitrate War. Chile declared war on both Peru and Bolivia on April 5, 1879. The Chileans were victorious and obtained all of the Atacama Desert area by the Treaty of Ancon, in 1883. The victory was of great economic value to Chile, for from 1879 to 1889 the duty on nitrate exports alone reached more than $557 million dollars, a very considerable sum in those days. The total value of nitrate exports exceeded $1.4 billion. It is noteworthy that the boundary question between Chile and Bolivia had existed for a long time but no one really cared about the matter until nitrate was discovered (in quantity) in the area.

Returning to how minerals are employed in warfare, in World War I the Germans made an interesting use of one particular metal which they had obtained from the United States. The Germans had developed the huge molybdenum deposit at Climax, Colorado, probably the largest single molybdenum deposit in the world; "Mount Moly," it has been called. These properties were, of course, expropriated during World War I. But the Germans had already taken quantities of molybdenum from "Mount Moly" before the outbreak of hostilities, for they were good metallurgists and knew that molybdenum made especially tough steel. They used this property of the metal in building the longest range cannon ever constructed. It was nick-named "Big Bertha" by the Allies and was used by the Germans to hurl shells into Paris from a distance of 75 miles.

The Germans knew that the physical damage from this sort of bombardment, from just one cannon, would be minor; but hoped that the shelling would have a devastating effect on the morale of the French. It did not, and was largely regarded as just a nuisance. But molybdenum still remains a critical metal in many steel products, both civilian and military. Without it, neither the ships and guns of the Navy, nor the tanks and guns of the Army could be built.

It early became clear to the German military that their operations were definitely dependent on having raw materials for the conduct of warfare and Germany was not self-sufficient in this regard. Accordingly, both to obtain markets for their finished industrial goods, and to obtain raw materials, chiefly minerals, Germany began to look at adjacent territories. After the Franco-Prussian War of 1871, Germany annexed the iron deposits of Alsace-Lorraine. Unfortunately, they later found that the boundaries they set did not include the bulk of the iron deposits, due to the nature of the geologic structure there. They apparently had a good military department but a poor geology department. So the First World War was fought, in part, by the Germans "to correct the error of 1871." Germany, however, by losing World War I, had to give back this territory.

From the experiences of World War I, both the Allies and the Central Powers (Germany and its allies) did become keenly aware of the need for minerals with which to conduct military operations. The slogan for the Allies became "never again," referring to some of the mineral supply problems they had experienced during the conflict. This triggered immediate and extensive post-war mineral exploration and development programs by Britain and France, for both foreign and domestic resources. As these countries at that time had extensive colonial holdings, especially Great Britain, there was much territory to explore. Germany, in contrast, had lost all of her foreign lands, which was, indeed, one of the circumstances that precipitated World War II.

Self-sufficiency, or as close to it as possible with regard to mineral and energy supplies, became one of the cornerstones of both British and French foreign policy. But by the late 1930s, Germany and Italy were demanding return of their colonies; and some military actions were commencing. Italy had invaded Ethiopia (called Abyssinia at the time); and Japan, also needing raw materials, was on parts of the Asiatic mainland, notably Manchuria, which they renamed Manchuko. Here, good deposits of two vital industrial minerals, iron and coal, were present in substantial quantities. In the late 1930s, Germany annexed Austria and was preparing to invade Czechoslovakia; and for the first time, the world was seeing a global series of political and military moves to obtain raw materials, chiefly minerals and energy minerals. The Industrial Revolution, rising populations, and a desire for a higher standard of living were the immediate causes, together with a background thought that should another war come, the materials must be available to survive and win it.

As Germany prepared for World War II, it was unable to take back its colonies, so it chose to annex its immediate neighbors, who might have some useful resources. Austria was next door and did have some small iron deposits and thus it was the first to be taken. Czechoslovakia contained some rather good and famous mining districts, with fairly large iron deposits; and a wide variety of other metals. Czechoslovakia was taken. With the early defeat of France in World War II — the Maginot Line proved very leaky — Germany again obtained control of the iron of Alsace-Lorraine.

Hitler then turned toward the east. He had long been of the view that only the Soviet Union had adequate land and minerals to take care of Germany's needs. He remarked that by taking this region "we shall become the most self-supporting state in every respect in the world. Timber we will have in abundance, iron in unlimited quantities, and the greatest manganese-ore mines in the world, and oil — we shall swim in it." So Hitler turned his armies east, toward the Urals and to the Ukraine. This area, along with the Donets Basin, collectively contain extensive deposits of hematite (high grade iron ore), excellent coking coal, and limestone — the fundamental three ingredients for steel-making. Germany, Hitler thought, was now on its way to becoming a permanent world power, and his troops headed into the Soviet Union. They were stopped at Stalingrad, and also at Leningrad, by heroic Russian defenders. The Russian winter also did its part.

It was not only metals that became so clearly vital as tools of warfare in World War I. Energy, chiefly in the form of that relative newcomer, oil, was obviously going to be very significant. Some of the ships of the Allies, bringing troops and supplies across the Atlantic from the United States to Europe, were coal-fired and some were oil-fired. The oil-powered ships were the better and faster vessels, greatly aiding in the more rapid delivery of both troops and equipment. Oil was becoming more important in warfare. Gasoline-powered tanks made an appearance on the military scene in World War I. Airplanes, primitive as they were, also came into the war in a limited fashion, powered by gasoline. Trucks, for the first time in a major conflict, replaced horse-drawn vehicles on a large scale. Seeing all this, world military establishments, in the time between World War I and World War II, began to give serious consideration to oil supplies. This is why Hitler wanted so much to move into the oil fields of the southern USSR.

During World War I, the British Empire was intact and as the Allies had control of the sea lanes, raw materials for warfare were funneled in from all parts of the globe. As World War II began to loom as a possibility, Great Britain became increasingly concerned about insuring the proper flow of war materials. The United States, also, at a rather late date, made a move (in 1939) when the U.S. Congress authorized funds to start an emergency stockpile of certain critical materials. Fortunately, Britain was still in greater or lesser contact with its possessions and former possessions around the world, and to a large extent still had control of the sea lanes, so materials could be brought into both the United States for processing into finished war material products, and also into Great Britain.

But with the total onset of World War II — both Japan and the German-Italian Axis becoming involved — the Japanese and German submarines made shipping lanes more hazardous. The situation at that time and during the rest of the war was summed up by Simon Strauss, who was one of the officers of the Metals Reserve Company, set up by the United States government, just before World War II, for this emergency. Strauss stated: "By the time Pearl Harbor broke out, all of Europe was under the domination of Hitler. Africa, Latin America, Australia, India, and even parts of China remained accessible to the U.S. The only competitive customer for the mineral exports of these vast areas was the United Kingdom. An Agency called the Combined Raw Materials Board was created by the U.S. and the U.K. and a coordinated buying program was launched. It was possible to fight World War II without an actual shortage of these critical materials. But the key was the fact that we retained access to Latin America, to all of Africa, to most of Asia, and to Australia. Had we been denied access to them, we would have been in trouble."

Thus the United States and Britain had access to large supplies of raw materials at the outset of World War II, but this was not true for the Japanese.

After Japan was opened to the rest of the world, in the late 1800s they gradually resolved to become an empire; but they had very few natural resources. The nearest substantial coal and iron deposits were in northern China (Manchuria), and became the initial target. The Japanese invaded Manchuria. Later they went into southeastern Asia, to obtain raw materials there, also. This continual Japanese expansion to obtain control of new territory — and therefore resources — ultimately brought the United States and Japan into conflict in World War II.

This over-riding concern by the Japanese for their vulnerability in the matter of energy and mineral resources is evident in a publication which came out after the war. The minutes of the Japanese General Staff meetings, just prior to Pearl Harbor, show the principal reason for their expansion was to correct, as far as possible, their deficiencies in raw materials. This involved the invasion of China and other parts of mainland Asia. The United States at the time had a Far East policy which, between 1931 and the time of Pearl Harbor (1941), supported the concept of "Open Door" trade and also supported China's independence from foreign invasion. But against the Japanese encroachment of China, the United States could only be indignant. However, this indignation rose to a peak in 1940. Up until that time, the Japanese had little fear of the United States, as the USA had a relatively small navy, which was split between two oceans. The Japanese, being a seafaring nation, had a relatively large navy, and it was only in one ocean.

The Japanese continued to expand their sphere of influence, largely through military means. The United States ultimately decided to invoke sanctions; and in July of 1940, put an embargo on aviation gasoline. But this was only a nuisance to Japan, for other grades of gasoline could be cracked in Japanese refineries to produce aviation gasoline. In September of 1940, the United States embargoed scrap iron and steel shipments to Japan. Up to this time, the United States had been chiefly concerned with matters in

Europe. Asia was a secondary consideration; but when the Tripartite Pact was signed in September, 1940 among Germany, Italy, and Japan, the shape of things to come began to emerge. With the signing of this pact, which said that each of the signatories would go to the aid of any other in time of war, the war which was already in progress in Europe since September 1, 1939, became fused with the Far East; and this made the United States more anti-Japanese than before.

But all was not well for the Japanese. They had hoped that the Germans would finish off Great Britain, thus opening up the areas of British influence in Southeast Asia to the Japanese. Instead, Germany declared war on the Soviet Union, a country with whom the Japanese wanted to have better relations. With this turn of events, the Japanese began to see more and more that they would have to go it alone; and so they increased their preparations for war with the United States, their only potential enemy of consequence in the Pacific region.

The minutes of the Japanese General Staff clearly show the great concern of the Japanese for their ability to survive a war with the resources on hand at the start of hostilities and those which they thought they could obtain immediately thereafter. The chief concern was oil. The United States and what was at that time the Dutch East Indies (now Indonesia) were the chief suppliers of petroleum products to Japan. In July 1941, Japanese troops entered southern Indo-China. Britain, the Dutch East Indies, and the United States immediately placed an embargo on all exports to Japan, including oil. This cut Japanese oil imports to about 10 percent of their previous volume, and immediately Japan saw itself with its storage tanks of oil (like an hour-glass) gradually running down. Something had to be done and done very soon.

The Japanese General Staff now had to make some decisions. If Japan were to retain its position as a substantial power, and maintain or improve its standard of living, it had to have access to raw materials. Without oil to fuel its merchant ships and its navy, this would be impossible. Japan would then have to gradually retreat back to its own islands and remain an island-bound, limited economy. The hour-glass of oil would last two years or less. Under conditions of intensive warfare, the oil might last only six months. "An outline plan for carrying out the national policy of the empire" was drawn up and presented to a meeting of the Liaison Council on September 3, 1941. Faced with this situation of either beginning to withdraw or trying to move ahead, it was decided to go to war with the United States, if an agreement on oil could not be reached by October. Oil was the critical matter. If Japan had possessed adequate oil deposits within its own borders, the Japanese would not have gone to war with the United States. Japan's lack of oil precipitated the attack on Pearl Harbor.

The record shows, however, that there was a split opinion among the Japanese leaders. Some thought that diplomacy could conceivably win the day, and therefore there was an increased effort in Washington on the part of the Japanese to change the position of the United States. This faction of the Japanese government tried very hard and up to the very last minute to achieve a change in U.S. policy, which accounts for why Japanese diplomats were in Washington at the very hour of the attack on Pearl Harbor. It was not because the Japanese were trying to be deceptive, as has been commonly thought in the United States; but it simply represented a genuine effort on the behalf of some of the Japanese to avoid war. But the United States, chiefly under the direction of Secretary of State Cordell Hull, stood firm and insisted that Japan essentially abandon all of its positions in Southeast Asia and Manchuria, and also other parts of China. The Japanese offered to simply withdraw from Southeast Asia, but Cordell Hull would have none of that, although President Roosevelt was said to have been at least mildly interested in the idea.

Faced with having to abandon all thoughts of a Japanese empire and being reduced to their island position alone, the Japanese attacked Pearl Harbor on December 7, 1941. The military had prevailed over diplomacy. Some say the Japanese had no choice. The key to the decision was oil. Without oil supplies, they were dependent on what they currently had in storage — those tanks were indeed big hour-glasses of time running out.

When the decision for war was made, and December 7th came, the Japanese had two inter-related immediate objectives. The first was to destroy, or at least neutralize, the United States fleet in the Pacific. The second was to immediately move south and seize the oil fields of the Dutch East Indies. To do this, the U.S. Fleet had to be sufficiently hurt so that it could not interfere to protect the Dutch East Indies. This is what happened and the Japanese did take the oil fields from the Dutch. If the United States Fleet had been able to intercept the Japanese, as they headed for the Dutch oil fields, the war probably would have been over in less than year. Oil was the key and the Japanese got their needed supplies, at least for awhile.

Whereas the war in the Pacific was triggered over oil supplies, and World War II in Europe was basically started over land, in general, and the resources contained therein, a metal which at the beginning of World War II was not much more than a laboratory curiosity ultimately ended the war. This metal is uranium, which before World War II did not have much of a market, and, in fact, was regarded as an undesirable impurity in some ores. Silver ores in the Coeur d'Alene mining district of northern Idaho, particularly those from the Sunshine Mine, contained some uranium and this was penalized at the smelter. Some of the rock from the mine, which contained uranium, was simply used to make and maintain part of the road to the mine.

But the fact that the United States possessed this metal allowed the USA to dramatically end World War II by means of the two atomic bombs that were dropped on Japan. Oil started the Pacific war and uranium ended it. It should be noted that oil also helped to end the war in Europe. Hitler's motorized divisions, toward the end of the war, suffered markedly from lack of oil. And when General George Patton was finally on the move across France, with the Germans in full retreat, pipeline specialists from Texas (where else!) followed Patton's tanks and laid pipe at the rate of up to 50 miles a day. And there was oil to fill those pipelines; for during World War II, the Allies controlled 86 percent of the world's oil.

And oil has continued to be a factor in warfare throughout the hostilities in the Persian Gulf (in the 1980s) between Iran and Iraq. In this instance, it was the ability to market the oil and thus obtain foreign exchange with which to buy the weapons of war that made oil so important. Iraq, greatly outnumbered in terms of fighting men and equipment, initially had a very difficult time holding its own against larger and more powerful Iran. The materials of war had to be purchased, for both sides, through the sale of their oil resources. Iran had the Kharg oil terminal, on the Persian Gulf, and also largely controlled access to the Gulf through which almost all of Iraq's export oil had been travelling. Initially, therefore, Iraq was at a great disadvantage in selling oil abroad to pay for the war. Iraq survived this early crisis because of massive aid from friendly Arab nations, who used their own oil to buy war materials for Iraq. (Iraq is both Muslim and Arab; Iran is Muslim but not Arab).

With the help chiefly from Saudi Arabia, Iraq eventually built a pipeline to the Mediterranean, followed shortly by a second line; and then a pipeline from Iraq to the Saudi Arabian port of Yanbu, on the Red Sea, which Iran could not control. By these means, Iraq was able to increase production from virtually nothing to more than a million and a half barrels a day. This turned the tide in the war — a clear example of how mineral resources can affect the future of nations through the procurement of vital foreign exchange. In this case, it literally meant the survival of Iraq.

Later, as Iraq became less vulnerable, it began to attack the Kharg oil terminal, rendering it at least partly useless as an oil outlet. Iran had no pipeline alternative to markets now, as did Iraq. As a result, Iran's ability to deliver oil to the markets was substantially reduced; and thus the initial advantage which Iran had over Iraq was lost and the war became an eventual stalemate. Furthermore, some popular support for the war was lost in Iran because without substantial oil money, Iran could no longer subsidize (as much as it had in the past) the large imports of foodstuffs (chiefly wheat and rice) which were distributed at reduced prices to the general population. The ability or lack thereof to market oil was the swing factor in this conflict.

Because Kuwait and Saudi Arabia both tended to support Iraq, Iran attacked oil shipping from those countries. The importance of the Persian Gulf oil (coming from Kuwait, as well as other sources) to the industrialized world was emphasized by the ultimate presence of naval forces of the United States, Great Britain, France, Italy, and the Soviet Union in the Gulf. A number of Kuwaiti tankers were re-flagged with the U.S. flag and were given naval escort through the Gulf. It was variously estimated that, given the cost of maintaining the naval presence in the Gulf, each barrel of Persian Gulf oil cost the United States about $135. The fact was, however, that most of this expensive oil went to western Europe and Japan.

We have been discussing minerals as they have related to their use rather directly in warfare or as a cause of war. But with the advent of the atomic bomb, it may be that global military action by the major powers is not so likely in the future, as it is probable that atomic warfare would result in everyone's losing. Therefore it seems more probable that actual warfare will be of a limited nature; and in the major battles of ideologies, the war will be *economic* warfare. The side which survives will be the side that can produce and maintain the most robust economy. The disintegration of the Communist economies in Eastern Europe, in the late 1980s, seems to be an example of this. Here again command of mineral resources is vital to winning that conflict. An aspect of this is control of economies other than your own by controlling resources vital to them. In this regard, oil and gas can be used in economic warfare.

The first use of petroleum in economic warfare developed when Great Britain tried to maintain control of the Suez Canal, in the 1950s; and Saudi Arabia cut off oil shipments to England. But at that time, the United States still had surplus oil producing capacity and simply opened the valves wider and supplied Great Britain's needs.

The Arabs lost that skirmish. But in the 1970s the Arabs and Iran discovered that the United States was no longer self-sufficient in oil. As a result, the Gulf nations were able to do two things: They could exert political influence on the United States (regarding the Israeli situation), and they also realized they could begin to take control of their own economic destinies and could begin to compete industrially with the western world.

From about 1940 to 1970, the price of Middle East crude oil had held between about $1.45 and $1.80 a barrel. This was in dollars, not adjusted for inflation, which meant that the price of oil had actually gone down during that period. But with the discovery that the western world, dominated by the United States, was no longer self-sufficient in oil, the Gulf countries, and the rest of OPEC, raised the price (in less than eight years) to more than $35 a barrel.

This was purely and simply a declaration of economic warfare on the west. And this warfare continues to the present time, in the form of industrial developments of some of the nations of the Persian Gulf, including the petrochemical complexes of Saudi Arabia and the United Arab Emirates, and the aluminum refining capacity of the latter, competing directly with similar facilities in the west. This economic aggression will continue in the future and will probably intensify, as the oil producing capabilities of the world become more and more concentrated in the Middle East, for geology has so determined that this will be the case.

The Soviet Union currently has the world's largest known gas reserves. The USSR is adjacent to western European industrialized countries, which have to import substantial energy supplies. Although the apparent immediate intentions of the Soviets are simply to sell natural gas to obtain foreign exchange, ultimately the result could be that western Europe would become more and more dependent on the Soviet Union for energy supplies — and potentially vulnerable to the influence of the Soviet Union. At the present time, the USSR supplies only about three percent of western Europe's energy, but a survey of total world uncommitted (surplus) natural gas supplies, as of 1988, showed that the rest of the world, not including the USSR, had 598 trillion cubic feet of surplus, whereas the Soviet Union, alone, held 763 trillion cubic feet. Obviously, over the longer term, the USSR will be in an increasingly strong position in the European energy supply situation, and could exert a corresponding influence on this area. For this reason, the construction of the large diameter gas pipeline from the gas-rich USSR Yamal

Peninsula into western Europe was vigorously opposed by the United States and an embargo was placed by the United States on equipment which would help to build the line. But this was a futile gesture, as similar equipment was available elsewhere. So the line was built and now brings natural gas into western Europe.

So far the amount is modest, but it is growing. The valves controlling the gas supply remain on the Russian side of the border; and little by little, western Europe is buying more and more Soviet gas.

Sweden has voted to phase out all existing nuclear plants eventually, and to build no new ones. As Sweden has no coal or oil and a modest amount of hydropower, where future energy supplies are coming from is a problem which seems to have been ignored, to some extent. Sweden has begun talks with the Soviet Union concerning possible purchase of Russian oil and gas supplies, this in spite of the fact that relations with the Soviets have been somewhat on the frosty side, due in part to repeated incursions into Swedish waters by Soviet submarines. In 1988, Sweden's OK Petroleum Company signed a contract with the USSR to purchase two million metric tons of crude oil and oil products. What the ultimate result of this decision by Sweden will be, only time will tell. But it would make them somewhat economically vulnerable to the Soviet Union.

Finland already gets some of its gas from the Soviet Union, through a pipeline into southeast Finland, completed in 1974. The pipeline network has now been extended into the Helsinki and Tampere districts. Seventy-three percent of the oil Finland uses now comes from the USSR. Finland also gets substantial coal from the Soviets. In regard to the Federal Republic of Germany (commonly referred to as West Germany), 15 percent of its energy is in the form of natural gas, and of this, 71 percent was imported — 24 percent coming from the Soviet Union. When these figures for West Germany are sorted out, the amount of energy obtained from USSR gas currently is not very great; but it is likely to grow, for the Soviets do have the resource and will have it in quantity for a long time to come. West Germany also imports 95 percent of its oil, about 6 percent of which now comes from the USSR.

The Turks have long been a thorn in the side of the Russians, because they control Russia's shipping and naval access to the warm waters of the Mediterranean, through the Dardanelles. And the border between the two countries has been a tense one, at times. But the Soviets now have penetrated energy-short Turkey with their natural gas lines; and in the Turkish capital of Ankara, Russian gas heats homes and factories.

The Soviets obtain about 60 percent of their hard currencies through the sale of oil and gas. Without this income, the Russians would have a difficult time supplying their war machines with needed technological equipment (computers are an important item). Oil and gas earnings also buy grain, something in chronic short supply in the Soviet Union. Grain bolsters the civilian economy, but grain also feeds the large armies which the Russians have. So, in a very real sense, oil and gas can be weapons of war during peacetime — economic warfare.

Poland is an example of a country which has, in effect, already lost this battle. Nominally it is an independent country, but its leaders are beholden to the Soviet Union for some basic reasons. The Soviets can bring Poland to an economic standstill merely by cutting off its oil supply. At the present time, the USSR exports about 80 million barrels of oil and about 19 million barrels of petroleum products to Poland, annually. Poland is also dependent on the Soviet Union for large supplies of fertilizer (made from natural gas), magnesium, and nickel.

Economic warfare was visualized by the Father of Communism, Lenin, when he wrote in 1916 about the possibilities of cutting off vital mineral resources from the western nations; and this view has been enunciated by subsequent Soviet leaders. Termed the "real war" by some political observers, this is a struggle which has many aspects. How consistently it is being used by the Soviet Union, in particular, is a matter of some debate; but the political stirrings (and in some cases, upheavals) in certain mineral rich African nations seem to have at least some communist involvement. How serious this is, only time will tell.

Probably more predictable, in terms of Soviet involvement in an area of great concern to the western democracies, is the growing Russian interest in the Persian Gulf, especially as the Soviets are now faced with a peaking out of their own domestic oil production, against a growing internal demand. The Persian Gulf region is one of marked and long-standing political upheavals. It is likely to continue so, and as it is in the Soviet's backyard, it is a fertile and convenient neighborhood in which to expand Soviet influence. And the fact remains that what is currently the most vital resource to the western democracies and (also to democratic Japan) comes from this historically and most unstable area. Former Russian foreign minister Molotov has stated that the Persian Gulf region is where lies the "center of the aspirations of the Soviet Union." Whether what has proven to be an unfortunate experience by the USSR, in moving southward toward the Persian Gulf, in Afghanistan, combined with the new leadership in the Soviet Union, will alter these aspirations remains to be seen.

The Soviets do have a formidable array of energy mineral and mineral resources within their own borders, which can be used in future economic warfare. They have tremendous reserves of natural gas, the world's largest deposits of coal, and still largely undeveloped copper resources, as well as large deposits of other vital minerals. Although Soviets are nominally atheists, I was told a story which included a reference to God when I visited the Geology Building at the great Siberian research city, a short distance south of Novosibersk. The guide said "When God flew over the world, holding the mineral resources in His hand, He opened it over Siberia."

To take advantage of this wealth has been difficult because of limited access. To correct this, the Soviets decided years ago to build another railroad through Siberia, to augment the heavily overloaded, original Trans-Siberian line. This new line, termed the Baykal-Amur-Magistral (the BAM), has recently been completed. It has opened up a mineral-rich region, and will allow the USSR in the future to export a variety of important resources. In turn, the Soviets will be in a position to influence world mineral markets on into the next century.

The USSR has been slow in developing its great natural resources, at a considerable disadvantage to the standard of living in the country; but by being late in such development, it also means that they still have these resources. In contrast, the United States developed its resources early and rapidly, which circumstances propelled the United States to the top of the world's living standards. But in the process, the cream of the energy mineral and mineral resources was skimmed off.

The Soviets still have much of that cream in their mineral deposits, and this will be a factor in economic warfare of the future. If the Soviets can really reform their economic system and make it comparably productive to those of the western democracies, this would have a tremendous impact on the western world. It could catapult the USSR into a position of world economic leadership, for they have the resources to do it, being the most nearly self-sufficient nation in the world in terms of energy mineral and mineral resources.

There has been tried at times an interesting variation of using minerals in economic warfare, whereby rather than a country with-holding a mineral resource, an importing country puts a ban on it, in an effort to influence, in some way, the producing country. An example is the embargo the United States at one time put on the importation of chromium from Rhodesia (now Zimbabwe), to pressure the then white Rhodesian government, with respect to its racial policies. But the United States still needed chromium, and the only other substantial supplier was the USSR. The Soviets promptly tripled the price of their chrome and what they could not supply themselves to the United States, they bought from Rhodesia, also tripling the price of the chrome from that source, as it was re-sold to the United States. The United States finally had to give up on the embargo.

As a footnote to that sad but educational experience, it is interesting to observe that the 1988 Congress of the United States passed a bill to ban United States investment in South Africa. This, again, as in the case of Rhodesia, was designed to influence the government's racial policies; but the Congress was careful to include an exemption for strategic minerals. They had learned!

Another variation on using minerals in economic warfare has been practiced by Saudi Arabia, which has on occasion driven down the price of oil by using some of its great oil-producing capacity to pump excess oil into the market place. This has been done in an effort to whip the other OPEC members into adhering to their production quotas, rather than cheating on them.

Economic warfare clearly exists in various forms today, and the drama will continue. How it plays out, no one can predict; but it is likely to continue playing for the indefinite future, and may intensify as resources become scarcer against rising industrial demand around the world — especially as less developed nations such as China become more industrialized, with a concurrent demand for more energy minerals and minerals. We can only hope that these needs can be accommodated without military action.

More recently, some disputes over minerals have been settled peaceably. In 1989, Australia and Indonesia, by negotiation settled a disagreement about ownership of certain waters off the Island of Timor. No one had really cared about this area until oil was struck there and then the matter of ownership was raised, but it was settled across a table.

Like the Australian/Timor question, the prospects of oil beneath the land have renewed a dispute between Venezuela and Guyana, over the Essequibo territory that is claimed by both countries. The Guyana Natural Resources Agency issued exploration permits to several oil companies for this region, but Venezuela disputes Guyana's right to the territory. As of 1989, however, this matter was still under peaceable discussion.

A number of years ago in the United States, oil was struck in the valley of the Red River, between Oklahoma and Texas. What had been the more or less accepted boundary between the two states then came into dispute. The precise legal boundary was the river at the time that Texas became a state, but the river had changed channels since then. The question was finally resolved by the aid of students of tree rings, who determined where the channel had been in 1845.

For all of these more recent peaceable approaches to settling disputes over mineral resources, there is, in fact, currently, a very real and shooting war of which very few people are aware. Sometimes called "the forgotten war," it has been going on for more than 12 years; and it is between Morocco and Algerian-backed forces, called the People's Front for the Liberation of Saguiat el Hamra and Rio de Oro, otherwise known as the Polisario Front. The actual battlefront is more than 1000 miles long, lined with rocks, minefields, and radar and artillery emplacements. The argument is over a 103,000 square mile area of the Sahara, wherein there are very large uranium-bearing phosphate deposits. A supporter of the Algerian-backed forces says that the region could be the "Kuwait of Africa," with a per-capita income of $16,000 annually from the phosphates, plus the coastal fishing potential. This conflict presents no apparent early resolution, but it is clearly a contemporary war over mineral resources. Can we ever finally leave such conflicts behind, or will they grow in the future as mineral resources become ever more scarce?

We may hope that all aspects of warfare involving minerals, both their possession and demands for them, and their use in more subtle economic warfare can be mitigated henceforth through a sense of world interdependence and relatively free trade in vital materials. However, in April, 1990, President Gorbachev of the Soviet Union ordered energy supplies drastically cut to keep Lithuania in line. Almost all of Lithuania's coal, oil, and gas, come from the USSR. It was raw economic warfare, using minerals as weapons.

Chapter 4

The Current War Between the States

One of the decisive advantages which the North held over the South in the Civil War was the possession of a large steel industry, based on the iron ores of the Lake Superior district. The South had to set up its own iron industry, based on what inferior iron ore they could find. Although no subsequent shooting wars have broken out between the states, the unequal distribution of minerals, and particularly energy minerals, combined with the great rise in oil prices during the 1970's, caused some hard feelings. The states, in the case of the United States, and also the provinces in Canada have been sharply divided between what has been called the "energy-producing areas, and everybody else."

We are talking chiefly about money which comes to a state or provincial treasury from royalties, and from various kinds of taxes, including severance taxes on oil, gas, coal, and other raw materials taken from the land of that state or province. If that state or province exports a substantial part of the oil, gas, coal, or whatever, then it is the out-of-state or out-of-province consumers who pay the government revenues derived from that resource.

The Province of Alberta is in the interesting position of producing about 85 percent of Canada's oil — a much greater percentage than the State of Texas produces in the United States. The result is that Alberta gets a very large amount of revenue from this source, some of which is paid for by oil-less provinces, such as Quebec. The annual per capita income in Alberta, from energy production alone, in 1981 was $1,000, while in oil-less Quebec it was $18 (from export of hydro-electric power). Quebec, however, has a large population and many votes. The result has been some hard fought, inter-provincial political battles.

Severance taxes which are levied by the states are usually calculated as a percentage of the value of the product; and, increasingly, energy exporters are putting severance taxes on their products. And it is the consumers in the energy-poor areas who have to pay this bill. The result is that the energy-rich areas export a substantial part of their tax burden to people in other regions. A study showed that the tax load on a family of four, with an assumed annual income of $17,500, ranged from 4.2% in Louisiana and Oklahoma, for state and local taxes, and Texas, where the family paid 4.5% of its income, to Michigan, where the family paid 9.6%, and to New York, where the family paid 11.5% of its income to state and local taxes.

The citizens of these various areas had no merit or lack of merit of their own to account for these inequalities. It is entirely the result of their geological luck, or lack thereof. Why did Louisiana have such a low tax burden? Chiefly because in the fiscal year 1980-1981, for example, Louisiana took in $1.19 billion dollars in oil and gas lease charges, royalties, and severance taxes, and because most of the petroleum produced was exported to other states and these charges went right along with the product to the out-of-state consumers. During that same time Maine, Vermont, Massachusetts, Florida, Wisconsin, and Idaho, to name a few of the energy "have-nots," didn't get a dime of income from oil and gas royalties or severance taxes. They just helped pay the bills for Louisiana.

The citizens of the great State of Texas would surely not look favorably on paying for the education of the school children in Minnesota. The fact is, however, that the residents of my native Minnesota help pay for the education of Texas children. This was determined some 400 million years ago on to about 20 million years ago. It was over that long interval of geologic time that Minnesota was only briefly invaded by a very shallow sea, which left virtually no appreciable thickness of marine sediments in which oil might be generated. In fact, most of that time Minnesota was busy shedding sediments southward toward the Gulf Coast, via the ancestral Mississippi River drainages. In the Gulf Coast area, great thicknesses of petroliferous sediments accumulated.

So, from the upper interior part of the continent came the sediments which were deposited over much of Texas and Louisiana, and on into the Gulf of Mexico — sediments into which organic material also was deposited that ultimately became oil and gas. From these petroliferous muds and sands, the citizens of these fortunate areas have had a very nice income, paid largely by the residents of less fortunate states, to whom the energy resources are sold.

In the early 1920's, the State of Texas set aside one section out of each township (36 sections) to be designated as "school lands"; and the income, whatever it might ultimately prove to be from such lands, would go to schools. Inevitably in Texas many of these lands proved to have petroleum beneath, and as this oil and gas is sold to oil-less Wisconsin and other energy-short areas, the income goes to support Texas schools.

So the State of Texas is really not all that unhappy about higher and higher oil prices. These high prices not only provide good royalty and severance tax income for the State; but, of course, high oil prices promote the oil industry and provide good jobs. Oil-less states, of course, would like to see low oil prices. Historically, Texas has produced about one-third of the oil in the United States. Royalties and severance taxes from that production have supported a friendly and comfortable economic environment over the years, which included no corporate or personal income taxes. Lack of these taxes also tends to encourage industry to move in, producing even more jobs and wealth, so the rich get richer and the poor . . .

All of this went largely unnoticed by the general public until OPEC boosted oil prices very high; and with Texas and other energy-producing states levying a percentage of the price of oil (or gas or coal), the treasuries of these states began to overflow with money, while the energy bills of the other states got higher and higher. From time to time, there has been a move by the federal government to try to limit the amount of severance taxes which could be levied by a given state, so as to more fairly distribute the wealth; but so far, the energy-rich states have wrapped themselves in the cloak of "states rights" and each state has remained its only little economic kingdom, in terms of minerals.

Various evidences of this "beggar thy less energy-fortunate neighbor" attitude appear, including such things as bumper stickers. On one Texas pick-up truck, I noted the statement "LET THE BASTARDS FREEZE IN THE DARK." Another favorite of some Texans, at least, was "DRIVE 90 AND FREEZE A YANKEE," referring to the fact the reduced highway speeds for cars would save fuel, and higher speeds would use more. But when the price of oil collapsed in the mid-1980's, the Yankees had long memories; and in Boston, Milwaukee, St. Louis, and Seattle, there wasn't much sympathy for Texas, with a corresponding abundance of Texas oil-man jokes, *e.g.*, "How do you contact an oilman in Houston?" Answer: "Call a cab." This joke referring to the fact that some oil-men who did go broke were actually driving taxis. These attitudes are unfortunate, for the United States is a single economy; and we all eventually, directly or indirectly, suffer together the hardships of any given area.

The oil price drop in the 1980s did hurt Texas to a considerable extent; and in 1987, they had to pass the largest tax increase (an increase in the sales tax, chiefly) that the state had ever enacted. But even then, there was still no personal nor corporate income tax and the average annual tax load per $1,000 in personal income for Texans was $53.76, whereas the average for the United States as a whole was $74.11. Oil revenues still pay a lot of the Texans' taxes;

and as most of the oil is shipped out of Texas, it is the citizens of the other states who continue to pay those taxes.

Under present laws, however, each state, in effect, is a nation unto itself, with a balance of payments situation involved with every other state. In terms of energy, there are currently 12 states of the United States which are energy self-sufficient and are energy exporters. These are Alaska, Kansas, Kentucky, Louisiana, Montana, New Mexico, North Dakota, Oklahoma, Texas, West Virginia, and Wyoming.

With the great price rise of oil in the 1970s (from around $3 a barrel to more than $35), there was nothing but joy in Houston and Midland, Texas, Tulsa, and Denver. The so-called energy crisis brought boom coal towns to Wyoming and Montana; and prosperity to many relatively isolated and previously poor communities in North Dakota, which were fortunate enough to be located in the oil-producing Williston Basin region. In 1981, with oil prices at more than $30 a barrel, Oklahoma enjoyed oil revenues of $100 million more than expected, with total severance taxes on coal, oil, and gas reaching $559 million. New Mexico, that year, passed the largest tax cut in its history and used taxes on energy (coal, oil, and gas) to slash income and property taxes. North Dakota used its taxes on oil and gas to take over funding of the schools from the local communities, markedly reducing property taxes. All these benefits were paid for by the energy-poor regions of the country, and represented — and continue to represent — a very considerable transfer of wealth, very much akin to how OPEC nations got rich, while the oil-less or oil-poor countries got poorer.

There are other examples of how the purely geological luck of some areas makes for vicious economic inequities. Stephen Brown, an economist with the Federal Reserve Bank of Dallas in 1990 estimated that with every $1 increase in oil prices, Texas adds 15,000 jobs and New York State loses 11,000 jobs, and the nation as a whole loses 78,000 jobs for every $1 rise in oil prices. Alaska, with Prudhoe Bay, ultimately had so much oil money that in 1981 it repealed its state income tax and did it retroactively back to 1979. Oil income for the State of Alaska in 1981 was about $10,000 per citizen. It was then decided that starting in 1982 the state would send each Alaskan citizen, man, woman, and child, a check each year, instead of having the Alaskan send the state income taxes. In 1982 each Alaskan received a check for $1000.

In one of my university geology classes, a year later, I asked if there was anyone present from Alaska. There was one student. I asked if he had received a check for $1000 and he said he had. I then asked who had paid for that, whereupon he rather sheepishly looked over the rest of the class and said "You did." The rest of the class was less than happy; for this was at the University of Oregon, and in oil-less Oregon that year, the legislature had to convene in special session to raise taxes to meet revenue needs, in part to pay rising costs due to the higher price of oil. For the oil which flowed so freely from Alaskan wells, and for which the average Alaskan citizen each received $1000 and did not move a finger to produce, Oregonians had to cut a lot of trees, make a lot of plywood, harvest tons of filberts, and catch thousands of salmon — all basic hard work — to earn money to pay for the oil they purchased from Alaska, and much of the oil used on the United States' west coast does come from Alaska. Some of that money was also transferred directly to each Alaskan, in the form of their annual oil income checks from the State. As of 1989, the checks were still coming. However, because of declining production and lower oil prices, the State of Alaska was a bit short of funds for this largess; so beginning that year, Alaska simply raised the oil severance tax to 15 percent, which, of course, is passed on to consumers in the ultimate price of gasoline and other oil products. As a final note, Alaska (in 1989) received 85% of the State's income from petroleum revenues. Since 1977, Alaska oil has provided $30.8 billion in State revenues; only $2.5 billion came from other sources.

"We in Montana, are wondering why our coal should light the streets of Seattle and Portland," so said the governor of Montana. But Montanans found a good reason "why" by imposing a 30% severance tax on Montana coal. That is, a 30% tax was added to the price of each ton of coal mined in Montana. Montana has a lot of coal. It does not have many residents and it does not have a state sales tax. The people who pay for the city lights of Seattle and Portland, as well as their own personal utility bills, are helping, in effect, to pay some of the taxes for the citizens of Montana. This is likely to continue for quite awhile, for Montana has very large coal reserves.

Montana and Wyoming are potentially the nation's OPEC of coal. In these two states lies more energy in the form of coal than exists in all the oil of Saudi Arabia. It is energy, but not in as convenient a form as oil, however; but sooner or later it will come into its own; and, like Saudi Arabia, Montana and Wyoming will rise in relative economic might. Both Montana and Wyoming are doing quite well from their mineral resources, although with the rather severe price swings in coal and oil, the picture changes from time to time. But with both states having relatively small populations, considering the size of the energy mineral resources which are being produced, the positive economic effect of these resources is quite large. And as the demand for energy increases, Wyoming and Montana will benefit correspondingly.

Attempting to reduce the high severance tax on coal in Montana has been the object of midwestern utilities which have to import this coal. Commonwealth Edison in Chicago estimated that the Montana tax costs its customers, annually, more than $14 million. With Montana's tax being so high, however, attention was focused in that direction for quite some time; and other areas, which had to buy Montana coal, complained loudly that Montana was getting rich at the expense of everyone else. Ultimately, because the complaints became so severe, Montana did reduce its severance tax to some extent.

Half of the Montana severance tax is put into a permanent trust fund, which will alleviate future problems caused by coal development. The other half is being spent for highways, schools, and other community projects.

Severance taxes on oil and gas in Texas are considerably lower than either Montana or Wyoming coal severance taxes; but Texas has had such large petroleum production, that severance taxes have produced as much as $3 billion annually. Currently, about seven states receive more than 20 percent of their revenues from severance taxes levied on the production of minerals, chiefly, coal, oil, and gas.

The Midwest Governors Conference estimated that residents of its 13 member states, in 1979, paid $700 million in severance taxes to other states; and that before the end of the century, the cost could be as much as $2.5 billion. As this runs up the costs for industry in these energy-poor states, it makes them less competitive in attracting new industry, than is the case with energy-rich states.

Whatever the merits of the argument, it is again an example of how the irregular geographic distribution of mineral resources can set one area against another — be it nations, or states, or even smaller communities.

The debate between the "have" and "have not" energy states can become quite heated at times. Congressman Sharp of Indiana stated "At some point I am convinced that the Congress will have to address the political, social, and economic issues raised by the spectre of a nation balkanized by inter-regional rivalry over energy and natural resource use and production." Congressman Tauke of Iowa noted that Iowa was paying more than $11 million a year to other states in severance taxes, and further stated that "Some states have now raised their severance taxes on coal to a level that cannot be justified by any reference to the costs that mining imposes on the state. Energy poor states cannot be expected to stand idly by while other states reap windfalls in state revenues — especially since these windfalls may be used to lure jobs from other states."

An editorial in *Business Week* magazine stated "Congress should certainly pass the bill limiting the tax. If anything 12.5% is too high a tax that is levied on consumers all over the country and contributes directly to inflation by raising costs. The Western states are abusing an ancient right and, in effect, taxing the citizens of other states."

"They are also gaining unfair advantage in the continuing contest to attract new industry and jobs. States such as Montana are putting the severance revenue into a trust fund but others are cutting their regular taxes and promoting themselves to business as super-low-cost areas."

"Montana and Wyoming control 70% of the nation's desirable low-sulfur coal. This country does not need a little OPEC astride the Rockies."

But there are some balancing factors in this war between the states on the benefits of energy and energy mineral resources possession. Exploiting these resources has some liabilities — mostly environmental — for a state. A state such as Louisiana may reap large economic rewards from its petroleum resources for a time, but it also pays a price.

California, which uses more oil than any other state in the Union, is not self-sufficient in petroleum. California owns the area from the coast line out for three miles and the Federal Government owns the offshore areas beyond the three-mile limit. From time to time, the Federal or State governments propose to lease these lands for oil drilling. But when drilling is proposed on either State or Federal lands, the hue and cry which goes up against such drilling is very loud. Californians do not want oil drilling to foul their waters, so basically what they are saying is "We love to have oil for our RV's, our cars, our beach buggies, and our boats; but don't get the oil from our area — drill in and foul up somebody else's backyard." And Californians are willing to pay the severance taxes on oil from other states, rather than produce the oil in their own area.

Those backyards are, among other places, the bayous and coastal marshes of Louisiana, and the bays and coastal areas of Texas. From the oil which these areas produce and sell to California,

Louisiana, and Texas do get considerable revenue; but there is a price, and this price is becoming more and more apparent. The result of oil drilling in the marshes of Louisiana, for example, has been to change the environment considerably, and almost always for the worse. To move in the big oil barges needed for drilling, canals have been cut in the marshes, which allow the invasion of salt water into relatively fresh-water areas; and the wildlife resources have been markedly and adversely impacted.

But the citizens of California do not seem to greatly care about the environmental problems of Texas and Louisiana, even though they vigorously proclaim concern for their own coastal waters. As long as the gasoline is available to fill the Los Angeles freeways with cars, where it comes from (as long as it is not from coastal California) does not seem to be of any particular concern.

One of my more vivid recollections is that of driving along the California coast one Labor Day. I came to an area where there were large numbers of cars and trailers parked. In the distance, over the ridge of beach dunes, I could hear a great dull roar. I stopped to see what it might be but I could see only one person there — a young lady standing on top of a sand dune, with a clipboard in her hand, on which there was a petition. I approached her and asked what it was all about, whereupon she asked if I would sign the petition. The petition was to prevent offshore oil drilling in California. (Parenthetically, it should be noted that it was not to bar offshore drilling in Texas or Louisiana).

I climbed up on the dune to discuss the matter with her, and when I reached the top of the dune, an incredible sight greeted me. Along the beach, there were literally hundreds of motorized beach vehicles, of all descriptions, roaring up and down the shore — back and forth, back and forth, in an aimless frenzy. I looked at the petition to prevent offshore oil drilling, and I looked at the myriad vehicles burning up fuel in mindless motion; and I said to the lady, "You cannot have it both ways; if you choose to pursue these activities, the matter of producing the oil should not be ignored or the problem passed on to other areas." She refused to speak to me after that. So, in effect, California, by refusing to allow offshore oil drilling to help supply its own needs, preferring to get it from Louisiana, Alaska, and other areas, is conducting environmental warfare, of sorts, against its oil suppliers.

In this regard, it should be noted that a considerable amount of west coast United States oil supplies, including those of California, comes from Alaska, chiefly the Prudhoe Bay field. As that field is now half empty, additional oil supplies from Alaska will have to obtained in areas east of Prudhoe Bay, which would include the Arctic National Wildlife Refuge. So, if California is to obtain more domestic oil, that is from where at least some of it must come.

There have been pressures brought to bear, however, on these provincial attitudes. At one time, a bill was introduced into the Congress which provided that California would get no more oil from the Four Corners region (Utah, Colorado, New Mexico, Arizona) until the residents of Santa Barbara stopped fighting the oil drilling in the Santa Barbara Channel. The bill did not pass but the message was clear — you should not dirty somebody else's backyard for a resource you can obtain in your own yard.

Actually, with present technology, offshore drilling is a relatively clean environmental operation. And both active and abandoned offshore platforms have greatly improved marine life habitat, locally, with, as a result, improved fishing. It has been the oil drilling in coastal marshlands, such as in Louisiana, which has damaged the environment.

But lest we seem to be picking on Californians, we note that this view of "get our resources somewhere else but don't disturb our environment" is a generally human attitude. Opposition to con-struction of oil refineries in the New England area (which needs lots of heating oil in winter) was the target of a bill introduced in 1974 by Senator Bellmon of Oklahoma, which would have prohibited the mandatory allocation of petroleum supplies to states which don't want to help produce or process petroleum. This bill died also. However, both pieces of legislation did serve to remind people that we are all in these energy and environmental problems together.

Offshore oil divides the states in other ways, too, and not just on environmental issues. Offshore oil obliterates party lines. Where there is money lying around, there are no Democrats or Republicans. It becomes a matter of region against region. In one example in 1985, it was a question of long-standing as to precisely how much money the Federal Government would get from offshore oil, and how much the adjacent state would get. The amount which had accumulated was about $6 billion — a big enough sum to arouse some very strong feelings.

The states with no oily coastal areas, such as Arizona and Wisconsin, were all for giving most of the money to the Federal Government. The states with oily coasts, such as Louisiana and Texas, wanted most of the money given to them. The fight was based strictly on geography, not on party lines. The states without offshore oil thought, rightly or wrongly, that the other more fortunate states were making out like bandits on royalties. The House Interior Committee reviewed the bill. The Chairman was Morris Udall, of Arizona (no coastline at all, much less an oily one), who made an impassioned speech against the bill. Committee members from Texas and Louisiana took the floor to say what a fine piece of legislation it was. The ultimate bill, which was passed, gave 27 percent of offshore royalties to the states and 73 percent to the Federal Government.

This law still did not give any Federal offshore oil money to inland states. In 1988, however, the states with no coastlines figured out a way to get at least some money from Federal offshore petroleum lands. The Supreme Court upheld a law which said that the individual states could levy a tax on that portion of an oil company's income which came from business within that state, and the oil company could not exclude from its taxable income the income it got from federal offshore oil production. Thus, as a result of this skirmishing among the states, the inland states finally got some offshore oil money; and oil money was what the warfare was all about.

What will be the future of these economic differences among the states and among the provinces, due to the unequal distribution of mineral resources? In the case of petroleum — oil and gas — those areas which have chiefly oil will be depleted first, and those with larger gas reserves will do well for some time longer, as gas reserves appear to be potentially longer-lived in the United States, at least, than do the oil reserves. But ultimately, as both oil and gas deposits are exhausted, other energy resources will have to take over, one of which is most likely to be coal, for coal is the largest single energy mineral resource in the United States, aside from uranium, in the breeder reactor form. The United States, in fact, has the second largest coal deposits in the world (USSR first), and probably the largest, in terms of economically recoverable coal. Montana and Wyoming, as already noted, are richly endowed with coal. Montana's oil and gas production is modest, and the potential there is not great for long-lived or increased petroleum production. But coal will be available in large quantities for many years to come. This is also true of Wyoming, where, however, at the moment, oil and gas are quite important. But eventually Wyoming and Montana indeed have the potential of becoming the OPEC of coal in the United States.

If the price of oil rises high enough, California may do somewhat better, for a time, as there are very large deposits of heavy oil in that state — oil which cannot now be produced by conventional means and takes special (and relatively expensive) methods to extract.

And ultimately we come to Colorado, Wyoming, and Utah oil shale, or the product of oil shale which is shale oil. Always, up to the present time, oil shale development of a commercial quantity, beyond the experimental stage, has been "just around the corner." If environmental problems can be solved and economics become favorable (much higher price for oil), Colorado may do fairly well with the extensive oil shale deposits of the Piceance Basin.

Utah and Wyoming also have modest oil shale resources. But the environmental problems are, unfortunately, very large, and revolve chiefly about water. The development of oil shale, one way or another, is likely to take a considerable amount of water and these deposits, for the most part, lie in the headwaters of the Colorado River — a stream which already has more claims upon it than it can meet. Beheading the Colorado River to develop oil shale would precipitate a huge cry from downstream areas, and would have a great negative environmental impact on those areas. It is a problem which at present seems to have no satisfactory resolution and may never be solved. Even before there has been any commercial development of this resource, the states downstream from the oil shale deposits have already protested the withdrawal of Colorado River water for shale oil production. Any such development would pit Colorado against California, Nevada, and especially Arizona.

In the case of Canada, the gods of energy clearly have favored Alberta above all other provinces. Alberta is now the chief conventional oil producer, but also is already producing some 200,000 barrels of oil daily by mining the Athabasca oil sands. As conventional oil and gas production declines, the tremendous Athabasca deposits will gradually come more and more on line, as will the more than 25 billion barrels of oil in the form of heavy oil in the Cold Lake region of eastern Alberta. Alberta also has considerable coal. So, Alberta will remain the Texas (or Saudi Arabia, if you choose) of Canada, for a long time to come. Quebec will fight for lower natural gas prices, while Alberta will vote for higher prices; and the Canadian government in Ottawa will to try to referee the argument.

In the United States, Texas may have to give way to Montana and Wyoming as energy producers, as oil production declines in the U.S. (Texas included) and coal rises in importance as an energy source.

The fact is that Texas' oil reserves have been declining rather rapidly in recent years, going from 14.8 billion barrels in 1960 to about seven billion barrels in 1988; and they are still declining. This trend is not likely to change, for Texas clearly has passed its peak as an oil producer. Still, Texas will be producing some oil for a long time to come, whereas Maine, Massachusetts, Minnesota and others never will produce any oil, but will have to continue to pay tribute to Texas.

Much as the "have not" areas would like to change the situation, it is probable that legislation can never economically equalize the "have" and "have not" areas of mineral resources, which were established by geological processes millions of years ago. Accordingly, some areas will always be poorer than others in terms of energy and energy minerals; but the relationships will change as one energy or mineral resource gives way to another, in a different area. The Mesabi Iron Range of Minnesota had its day, and ore revenues which that area collected were paid by all the steel users in the United States, wherever the iron end product was shipped. This tribute paid for many improvements in the Iron Range, giving it, for a time at least, a considerable economic advantage over the rest of Minnesota; and there was some envy expressed.

But the envy gradually subsided as the higher grade ores ran out and affluence slowly turned to depression. There was a war, economically; and the Minnesota iron range partially lost it, as the high grade hematite ore was depleted, leaving only lower grade taconite. Other places had better ore. This was inevitable, as it is with all mining areas, for sooner or later the ore deposits are exhausted. The same was true of the copper range of Michigan, on the Keweenaw Peninsula, which lost the war to Utah and Arizona, which had greater reserves. But these areas, too, are gradually losing the economic war with other parts of the world, notably Chile and Zambia, where higher grade ores and much lower labor costs exist, making Utah and Arizona copper deposits only marginally economic, except at times of exceptionally high prices.

So we try to live in harmony with one another, but beneath the very thin veneer, which we call civilization, there are basic needs of society which will be supplied by whomever has the resources — and, more particularly, the lowest cost resources. Frictions and economic inequalities will arise as some areas will have the resources and some will not. In the case of the United States and Canada, the unequal distribution of these basic resources will continue to cause regional envy, dissension, and hardships. And this economic warfare will probably persist, in one form or another, for as long as the resources exist.

Chapter 5

Minerals and Money, and the "Petro-currencies"

This chapter might well have been titled "Minerals as Money," for during many centuries the two were one and the same, and still are in a few places. Gold and silver, and sometimes copper, lead, tin, and zinc were literally used as money. In the eyes of many people, gold and silver still remain the only valid forms of money. And as long as politicians run countries, there will remain good reasons for that view.

Money, to be truly money in the classic sense, has to be a store of value whereby one person's goods or services can be traded for money and that money later given for goods or services of equal value. If inflation intervenes between these two transactions, paper money tends not to retain its value. On the other hand, gold and silver have proven to be much more dependable in that regard. Paper money, at times, has actually seen its value decrease by half or more in just a matter of days, as in the German inflation of the early 1920s. In Brazil, Peru, Chile, and a number of other nations, inflation has in recent times exceeded 100 percent a year. Next door to the United States, in just a few years, the Mexican peso went from 12 to the dollar to more than 2700 to the dollar.

It takes a certain amount of energy and materials to wring gold, silver, copper, and zinc from the Earth. It takes much less energy to cut down a tree and process it into paper and then print some figure on the paper and call it money. Also, unlike a gold coin, which has a fixed gold content (one ounce, for example) a piece of paper can be printed in any denomination. You cannot legitimately stamp any other figure than "one ounce" on a one ounce gold coin; but governments can and do print any number they wish on a piece of paper and call it money. Johannes Gutenberg, with his invention of the printing press, in effect also invented inflation; and every politician who votes for deficit spending should have a statue and a shrine to Gutenberg in his office.

The first use of metal for coins is lost in antiquity; but it may have started in China, as near as history can tell (a likely possibility, as the Chinese seem to have been first in a great many things). Interestingly enough, very shortly after the Chinese started issuing precious metal coins, other Chinese began to make them wholly or partially out of lead, so it may be also that the Chinese can be given credit for the lead nickel, and counterfeiting in general. This counterfeiting became so rampant that Chinese officials ultimately decreed that anyone operating a lead mine, without government sanction, would be executed.

The use of metal coins continued for many centuries, and the capture of state treasuries (such as the Persian, Babylonian, and Greek) were major military objectives. Along with slaves, the gold and silver booty was paraded down the streets of the victors.

The parade of the precious metals and the slaves vividly illustrated the importance of two things which are still vital for civilization — mineral and energy resources, the slaves representing, at that time, energy in useful form.

After Greece, Rome gradually rose to become the dominant power in the Mediterranean area. Rome did this by developing a large and well disciplined army. The soldiers were given a ration of salt as part of the monthly payment for their services, and from this we have the expression "is he worth his salt?" The fact that payment was part in salt in regular installments also gave rise to the term "salary" derived from the Latin word *sal*, for salt.

The Italian Peninsula itself has a fine climate but it is grossly deficient in precious metal. So to pay the army with something in addition to salt, Rome needed sources of gold and silver. Some had been obtained from looting of the Greek treasuries, but that supply ultimately dwindled. However, when Rome defeated Carthage, among the spoils of the campaign was an area known as Spania, now called Spain. From Spain, the Roman military returned with much gold and silver, as well as control of the mines in Spain which produced these metals. Silver ultimately became the main metal of Roman coinage, and the unit of exchange was the *denarius*. Rome, as noted, had very little in the way of precious metals, so it was the mines of Spain, as well as the earlier captured Greek gold and silver, which financed Rome's armies and made Rome the dominant power in the then known civilized western world.

The Romans, on their delightful peninsula, had luxurious villas overlooking the various bays on the Mediterranean. And they also had beautiful homes across from Italy on the south shore of the Mediterranean, for they had taken over parts of North Africa, including what is now Libya. The Romans loved luxury, and the gold and silver captured by their armies; and that produced from mines worked by slaves in the conquered lands, chiefly Spain, provided money for imports which Rome needed. Paper money had not been invented, so Rome could not run a deficit in balance of trade, as does the United States today. Precious metals were the only medium of exchange. Ultimately, precious metals from the mines in Spain became the chief source of money for Rome; and this situation continued for many years.

But gradually a difficulty arose. The mines got deeper and deeper, and the problem of water flooding into the mines became more and more difficult to handle. The Romans were quite ingenious for a time in regard to this problem; but ultimately they did not have adequate systems or pumps with which to remove the water. Finally, the ore which could be reached in the mines was exhausted.

There were many causes for the fall of the Roman Empire, but one surely was the demise of the Roman currency. The great gold mines of Spain, for three centuries, had produced more than 300,000 ounces annually. With the depletion of the mines, however, Rome's treasury became empty. But Rome had to continue issuing coins to pay its army, for the army had become, in its later stages, an entirely mercenary force. The sons of Rome were no longer drafted, as in earlier times. They chose to hire others to take their place, rather than leave the luxuries of Roman living for a spartan military routine.

The mercenary army's loyalty was *purchased*, and not in-born, as it would be with native sons; and the army definitely wanted to be paid in something of value. The Roman answer to the of lack of silver and gold for their coins was to debase the coins by adding other metals — chiefly lead and copper. (This also, by the way, is the recent story of coinage in the United States, where copper and zinc have replaced silver). But the citizenry and the army were not deceived. If you visit the museum at Sabratha, the ancient Roman city, a short distance west of modern Tripoli in Libya, you will see an interesting display of Roman coins. The earlier coins were pure silver and gold. But later coins, although carrying the same stamped face value on them, were debased. However, the citizens, the merchants, and the army knew about the cheapening of the coins. To compensate for this debasement of their money, they filed down the earlier coins to, in some cases, merely narrow, "pie-slice" shaped pieces to equal the value of the later issued debased coins. So on the wall where these coins are mounted, you see narrow slices of earlier pure precious metal coins, which remained in circulation and which were equal in value in that sliver-form to the value of the later, still-round but debased, coins.

The earlier Roman coinage, with its pure precious metal content, was the most highly prized and valued currency of its time in the Mediterranean. However, as precious metal supplies dwindled and debasement of the coins proceeded, but with the government continuing to pass these debased coins as full value, the Roman currency became a despised currency and with this change of view, the prestige of the Roman Government also declined.

Eventually, by the fourth century A.D., the situation was so bad with coins of various degrees of debasement in circulation, that it was necessary to weigh and assay each coin. In these circumstances, the Roman currency became virtually worthless. The mercenary Roman legions could not be properly paid, and their loyalty to Rome disappeared with the disappearance of the silver and gold from the coins. Also, the Roman general citizenry could no longer import the luxuries which they had once enjoyed. Finally, with the northern hordes beating on their doors and with no money of value to pay for defense, the Roman Empire collapsed. The debasement of the currency was actually only one of the reasons for the fall of Rome, but it was an important one. It may possibly have been the single most important one, for the Roman historian, Antonius Augustus, wrote "Money had more to do with the distemper of the Roman Empire than the Huns or the Vandals."

Ultimately, as we have noted, Herr Gutenberg invented the printing press; and gradually around the world, paper money has taken the place of gold and silver. Some countries still cling in various ways to precious metals for their currencies, chiefly by issuing special coins for various occasions. But in general, most countries (including the major nations) issue paper money which may or may not be backed by gold or silver — much of it is not.

If countries were forced back to the gold standard, it would speedily show what was happening to their currencies. The honesty of gold and silver as money cannot be tolerated by the politicians who, by voting for generous public spending, are able to use the public treasury as a campaign fund.

And so it was that in the 1930s, the United States went off the gold standard; and in the 1960s, the United States took all the silver out of the coinage. The silver coins were becoming more valuable, in terms of the inflated paper currency, than their stamped face value. Gresham's Law took over — "bad money drives out good money." Such silver coins as existed at the time speedily were taken out of circulation, as people searched their change for these bits of real money. The new coinage was nickel and copper — and later even the copper in pennies was replaced with zinc. Try to find a silver coin in circulation (in the United States) today. With copper coated zinc pennies, how far down can the debasement of a metal currency be carried?

Ultimately, it may be that even zinc is too valuable for pennies. Then what? The end is paper; and in Hong Kong, when you exchange your money for the local currency, if the final exchange figure has some pennies in it, you will get slips of paper — which are Hong Kong pennies.

The record of paper money has been miserable, as witness the 1923 50 million mark German banknote I have before me. In late 1923, near the end of the wild German inflation, workers were paid twice a day; and their wives met them at the plant gate at noon to get the *morning* pay, so it could be spent immediately because prices would be higher by evening.

But other things besides gold and silver can and do, in effect, back a nation's currency, to some extent. The ability to produce food, cars, refrigerators, medicines, medical equipment, and many other items tends to support a nation's money and therefore the materials from which these things are made, including land, iron, and copper; and energy resources in the ground (coal, oil, gas) are valuable. In the case of Japan, which has very few natural resources, the yen has been strong because of what the Japanese can do in the way of processing and up-grading the natural resources they import. But their economy is a precarious one, for without these basic raw materials within their own borders, they are vulnerable, and they know it. They are trying to compensate for this situation, at least in part, by buying into mineral resources abroad, either by 100% ownership or, more often, through joint ventures.

There is no better example of how the possession of mineral and energy mineral resources affects a nation's money than the case of oil. As oil became more and more important in world economies and clearly was the only thing making the money of some countries valuable in world trade (Nigeria, Qatar, United Arab Emirates, Kuwait, Libya, and others), the concept of "petro-currencies" developed. The term has been in use only fairly recently; but when Kuwait's chief resource lies beneath its desert sands, an amount several times greater than that remaining in the United States, there can hardly be a better description of the Kuwait *dinar*.

In the Persian Gulf states in total, oil provides about 80 percent of government revenues, and the sale of oil is the chief thing which supports these currencies. In North Africa, Libya has virtually no other resources than oil; and the Libyan pound floats on the surface of an ocean of world currencies, because it floats on oil. When the oil is gone, the Libyan pound will sink.

The same is true to some degree for the British pound, since the discovery of oil in the North Sea. With that discovery and the subsequent self-sufficiency of Great Britain in terms of oil, and actually, for a brief time, the possession of an export surplus of oil, the British pound has done a great deal better than in times past. It has had its problems, however, which were clearly tied to oil. In the 1980's, when the price of oil declined markedly for a time, the British pound dropped to the point where it was almost "a dollar a pound." Then, as oil prices rose, so did the pound — a clear example of the importance of that energy mineral resource to the value of British currency.

British Prime Minister Margaret Thatcher's greatest stroke of good luck was probably the coincidence of her early tenure with the surplus foreign exchange account, which the government ran on the basis of oil production and export. This has helped Mrs. Thatcher become the Prime Minister with the longest tenure during recent British history. Without North Sea oil revenues, the government's budget deficit would have been much higher; and if the North Sea oil income had not been available to reduce the budget deficit, it would have required income tax increases of between $30 and $35 billion. There would have been another million people added to Britain's jobless lines, which circumstance would have probably contributed to a considerable amount of civil strife.

Over the longer term, the North Sea oil has given Britain a lasting legacy in the form of some 80 billion pounds of foreign investments, which Britain made at the time it had the oil money with which to do it. This was earned as new oil fields were discovered and oil prices went up and up. As oil prices slumped in the mid-1980s, combined with a levelling off of production, the income from oil declined to the point where, in 1986, the income from investments made earlier from oil money exceeded the income directly received from oil. And this again tended to support the British currency and economy. Thus the British pound also has recently been floating largely on a pool of oil.

However, Britain has come close to killing the goose that has been laying the golden, or in this case, the oil-filled egg. The average tax rate, when adjusted for inflation, is about 78 percent of the profits of the oil companies; and for some oil fields, the rate is over 90 percent. With such a high tax, companies have declined to develop the many small field discoveries which have been made, with the result that Britain may be passing the peak of its oil production by the early 1990's. The only hope for further development lies in the more than 40 small fields, some with only one-twentieth the recoverable oil found in the larger oil fields. As one observer has said, "The government milked hell out of this industry, and the cow nearly fell down." But more recently the British Government has reduced its tax rate on all new fields, with the result that some of the smaller discoveries are now being developed.

In the late 1980s, the value of the U.S. dollar plunged to a record post-war low against other major currencies. The chief reason for this was the huge deficit in balance of payments, combined with a huge budget deficit at home. A significant part of the trade deficit resulted from the cost of importing oil and other raw materials. The oil bill alone was more than $100 million a *day*, nearly $50 billion a year and going up. It was and remains the largest single item in the U.S. import bill. It is likely to continue so as more and more oil has to be imported, to take up the gap caused by declining domestic production.

How sensitive the dollar is to oil was illustrated when the rumor came out that a major exploration project had struck oil in the Baltimore Canyon area, off the east coast of the United States. This had long been a major exploration hope and target. Upon the oil discovery rumor, the dollar immediately jumped in value. Subsequently, when the rumor proved false (and the Baltimore Canyon area has now been one of the great disappointments in east coast off-shore drilling), the dollar dropped.

The dollar continued to drop due to a variety of factors; and politicians have applauded this situation, for presumably it has made exports more competitive. But this low dollar situation is a two-edged sword. The weakening of the British pound (propped up temporarily by the oil from the North Sea) coincides with the decline of Britain's position as a world economic power. The pound, once the world's premier currency and the international unit of exchange, is no longer that. If, in turn, the United States' dollar becomes a greatly weakened currency, the prestige of the United States is also weakened. No one respects a country where it takes a fistful of money to buy a loaf of bread. The value of its currency, how it is respected, is to a considerable extent a measure of that country's political strength and influence around the world.

Oil is currently priced in the international market in U.S. dollars; but because of the continued weakness in the dollar, the oil producing countries have suggested that perhaps oil should be priced in terms of a "basket" of currencies — that is, by using the average value of several combined currencies. If oil indeed is no longer priced in U.S. dollars, it will be a blow to U.S. prestige.

Whereas the huge oil import bill for the United States is not the only culprit in destroying the image of the dollar as a strong currency, it is an important one. This factor is still growing and there is no early end in sight for this trend, as U.S. oil production continues to decline and imports correspondingly are going up.

In a more limited sense, state by state, for example, oil also is money in the form of how bonds are rated and how easily they can be sold. A brokerage house titles one of their investment publications "General Obligation Credits in the Oil Patch." The gist of this is that as oil goes, so go the economies and budgets of states such as Louisiana, Texas, Oklahoma, New Mexico, and, most notably, Alaska. In the first three states mentioned, bond ratings were lowered when the price of oil dropped and oil production and exploration activity were reduced.

In the case of Alaska, however, because of Prudhoe Bay's strong production, Alaska's bond ratings were raised in 1980, and have not been reduced as of this date. But when Prudhoe Bay production begins to dry up and if other production is not found in Alaska, no doubt that state's bond ratings will be reviewed. Alaska, more than any other state, has its budget tied directly to the petroleum industry. Revenues from oil and gas make up about 86 percent of the total unrestricted revenues of that state.

Looking abroad again, other countries aside from Britain and the Persian Gulf nations are heavily dependent on oil for maintaining the value of their currencies. In general, among the OPEC nations (some of which are not in the Persian Gulf region), between 50 and 90 percent of government budgets have been financed recently by petroleum revenues. Nigeria, Africa's most populous nation, has its economy now heavily built on oil. Without oil, Nigerian money would be of considerably less value in the international market place.

Oil, as we have noted, is currently priced and paid for internationally in dollars. As long as the dollar is strong enough to be wanted around the world, the presence of the dollar in a nation's accounts makes the budget easier to handle by providing universally acceptable foreign exchange and reduces the pressure on the local currency to carry the load. If the oil money were not there, it almost certainly would take a lot more of the local currency to buy anything on the international market. Oil, being priced in dollars which are accepted internationally, gives each oil producing country, no matter the condition of its local currency, money which is recognized as money around the world. Oil thus is indeed money to each such country.

There is also another aspect of oil income. Various political and social structures tend to be built up in some countries financed mainly on oil revenues. A variety of public works projects may be started, based on current oil income. If oil is not available to keep these programs properly lubricated with money, the people involved may become rather restless. It is, therefore, important to those in power that oil income continues to flow in; and this is one of the chief reasons why, from time to time, including at present, a certain amount of cheating takes place against the oil production quotas which OPEC desires to set for its members.

One might pursue this matter of oil and other mineral money a bit further by theorizing about what may happen as oil production naturally declines and so do revenues. What happens to the stability of a country built up on oil or metal revenues, when these monies are no longer available? Are economies and populations expanding out on a limb that will ultimately be sawed off? This is a very vital question and we shall shortly begin to see such changes. In fact, it is already happening in some countries such as Venezuela, where oil revenues peaked out several years ago. Venezuela has other resources to a degree, and large deposits of very heavy oil, which may ultimately be brought into substantial production; but for the moment, the in-place oil business is declining. Single resource countries such as Kuwait and the United Arab Emirates will have to survive profound changes in their way of life if they cannot (by judicious investment) put in place something to replace oil for revenue. In some cases this is being done, or lacking some other enterprises to develop within the country, they are simply investing their money abroad in areas which have more permanent and stable economies. This is further considered in Chapter 8 — *Mineral Riches And How They Are Spent.*

Oil money is not only aiding these Middle East countries that have the oil; but substantial populations in that general region are also becoming dependent on oil money, although those people may not have much oil themselves. This increasing dependency on an exhaustible resource, by expanding populations, is a potentially explosive situation.

Egypt, with its 60 million people, and growing rapidly, is increasingly dependent for its survival on foreign aid, much of it coming from neighboring, friendly, oil-rich Arab countries, chief of which are Saudi Arabia and Kuwait. But when the oil revenues of its rich neighbors begin to decline, Egypt can no longer expect to be supported, as at present and the recent past. With its population increasing by about one million persons every nine months, then what? There is no apparent happy solution. Is Egypt building out on an oil money limb which is not even their own and which will ultimately be cut off? In nature, the over-extension of a population upon a resource which diminishes is well known, and the results tend to be disastrous. Perhaps this situation will not be played out for some time, however, as both Kuwait and Saudi Arabia have very large oil reserves. But as the oil begins to run out in these two countries, they will no doubt look to their own interests first.

Another oil-less Arab state benefiting from its oily neighbors is Jordan. It gets a little export income from one mineral, some small phosphate deposits; but it has become a gold trading center for some of the Arab world. On the streets of downtown Amman, the Jordanian capital, may be seen robed Arab men and women frequenting the gold shops. Arabs have long placed their faith in gold; and as they have gotten wealthy on oil, they continue to pursue the faith to a considerable degree, perhaps rightly not trusting any of the currencies of the world. The gold trade in Amman is brisk.

Foreign exchange is urgently needed by many smaller countries whose economies are not big enough to justify the presence of an automobile plant, an electronics complex, laboratory facilities for the production of vital medical supplies, or other sorts of manufacturing facilities which must have a reasonably large domestic market to survive. Without a diversified industrial base, foreign exchange has to take up the gap and pay for the imported items. Where oil is not available as money, metals and other minerals may provide the foreign exchange so urgently needed. Peru, Bolivia, Chile, Ecuador, Libya, and Zambia, for example,have no automobile plants. All cars and trucks have to be imported. These countries are not highly developed technologically, and therefore cannot sell technology; nor do they have a manufacturing economy, such as does Japan, which, in spite of few raw materials, currently does very well. Accordingly, these lesser developed countries must fall back upon production of raw materials, commonly minerals, for their foreign exchange. These minerals, then, are the only generally accepted money which they have in the world market.

Bolivian tin accounts for between 40 and 70 percent of that country's foreign exchange, the exact amount depending, to a considerable extent, on the rather widely fluctuating tin market. Mexico obtains about 70 percent of its foreign exchange from the sale of oil. Peru gets most of its foreign exchange from exports of gold, silver, copper, and a little oil. Ecuador obtains more than half of its foreign exchange from oil exports. Zambia's copper provides more than 80 percent of its foreign exchange. Chile exports copper.

The dire need to export "something" for foreign exchange tends also to disrupt mineral resource markets. Chile, with its relatively rich copper ores and low-cost labor, can produce that red metal at an average current cost of less than 50 cents a pound, substantially below production cost in the United States. And even if Chile had to sell its copper at a loss, it would tend to do so, as it needs foreign exchange at almost any price. Chile at times has dumped large amounts of copper on the market, despite low prices and demands; the result in the United States has been that many copper mines had to close. But on the other hand, the money Chile earns goes in part to pay the interest on the $20 billion foreign debt owed by Chile, much of it held by U.S. banks. It is a complicated financial world which mineral resource economics creates.

Another aspect of this dependence by some countries, on the export of minerals or energy minerals, is the fact that as the price of these materials drops, there is need to export *more* to achieve the same level of income. In turn, by exporting more, still lower prices result, and so on. These countries need to export minerals not only to obtain foreign exchange, but also to provide internal employment. However, in having to produce more, because of declining prices, to obtain the same amount of foreign exchange, there is a compensatory factor in that more employment is created. This has the advantage of keeping the local population reasonably contented and is good for the political machine in power. Even if the price of the product falls below the cost of production in a given country, that country is still likely to continue production. This, at first glance, does not seem logical; but the economics of this was neatly explained by a Minister of Mines of one African nation, who said, in regard to a gold mining project which was by most standards uneconomic: "You do not understand. As long as the metal brings in more dollars than we have to spend for supplies from the outside, it is economic. The labor, power, and other supplies are obtained locally and paid for in our currency, the value of which we control. We need the thousands of jobs to keep the people quiet. Therefore the project is economic."

Thus it is not just the matter of earning foreign exchange which causes over-production and a decline in the price of minerals and energy minerals, but it is politics. The amount of the resource produced is increased in order to "keep things going." The "things" in many cases being the current political set-up, which may involve numerous channels of graft, keeping the present politicians in power. If these channels are not kept properly filled with money, there are likely to be changes and the winners will not be the current rulers. It might even come to the point where heads may literally roll. Some of the cheating on OPEC production quotas may have its origin in these circumstances.

But minerals, oil in particular, have had a much more profound effect upon the world's economic system and its money and its structure than simply influencing the value of a particular currency or providing a foreign exchange medium. When the price of oil was raised from less than $3 a barrel to more than $35, the world saw the biggest transfer of wealth in all history, and it continues. Being unable to locally absorb this flood of money (denominated in U.S. dollars, as the price of oil is, world-wide), the money from the oil-rich Arab world had to be deposited someplace where it would earn income until the time it could perhaps be effectively used internally in the country, in various projects. The money had to be "parked" somewhere and the chief place it was put was in the major banks of Europe and particularly the large banks of the United States.

Faced with this huge influx of money, these banks in turn needed to put the money out somewhere, so that it could earn what interest the banks had to pay on it, plus a profit for the bank. The places which needed money the most and apparently were willing to pay for it, or so they said, were the developing countries, which included chiefly Mexico, Brazil, Peru, Argentina, and several African nations.

The Arab money poured into the banks and the banks in turn poured it out to undeveloped regions. It seemed to be a happy arrangement all around.

But apparently the bank managers, in general, did not take time to realistically look at how these debtor countries might pay back the money, given their social, economic, and political situations, along with (at least locally, if not nationally) the dishonesty of some government officials — almost a tradition in certain areas. Some of these loaned dollars were subsequently siphoned off by graft and corruption. Some money escaped the plunder of the times and did get into projects of various sorts, many of which, however, were poorly managed, at best. Some or them were bad ideas to start with. Peru was one of the first countries to begin delaying debt repayment, with one of their reasons being "these projects were so bad you should have known better than to lend us the money we asked for on them." (This is quite literally what was said!)

73

Mexico, with a population of some 80 million people, was one of the first to line up to borrow from this pile of oil money in the U.S. banks. And Mexico now has built up a debt of more than $100 billion, which means that each Mexican man, woman, and child owes a foreign debt of more than $1000 each.

In a country where the annual income is about $700, clearly this debt is neither manageable nor repayable. The prospects that this debt will ever be paid off are almost nil. But to prevent the debt from appearing as a loss on the balance sheets of U.S. banks, the banks have been loaning out more money in order for Mexico to pay the interest on the debt — by means of this loaned money. The ultimate result, of course, is simply to make the debt even larger. In some circles, this would be regarded as putting good money after bad; yet in what appears to be rather weird international banking practices, this procedure seems to have become accepted. But this ostrich-like "bury the head in the sand" policy is not likely to make the problem go away.

What is particularly fearsome is the fact that the banks, in many cases, have loaned out more than their capital. And here we pause to make an elementary accounting point. The optimists with the ostrich posture say that the banks have loaned out only a percentage of their assets. The realists point out that the banks have loaned out a percentage of their capital. There is a vast difference between these two situations. Assets include all the loans the bank has outstanding, for presumably these loans will come back, paid in full. But most of the money loaned out is depositors' money. If the depositors should withdraw all their money, what is left is the bank's capital. This is the money put up by the individuals who started the bank. If the depositors take out all their funds, it is only the bank's capital which is left to pay off the bad loans.

On this basis, there are some interesting comparisons. The optimists point out that Manufacturer's Hanover Trust has loaned out about 10 percent of its assets to Brazil, Mexico, Venezuela, and Argentina. The realist will counter that this amounts to about 219 percent of the bank's capital. Chase Manhattan loaned out about 212 percent of its capital, and Citicorp loaned out about 203 percent. If these loans are not paid back, the banks face some very severe problems. It may be that the U.S. government will step in and assume some of the loans, which means, ultimately, that the U.S. taxpayer will be sent the bill. It should also be clearly under- stood that all debts have to be paid — either by those who initially borrowed the money, or by those who have loaned the money by losing it. But the debts will be repaid by one process or the other.

Once on the books, the debt has to be accounted for, one way or the other.

Thus, in the reorganization and also the destabilization of world economies, and particularly the banking systems, which oil has brought about, not only may it prove to be that the United States will find that it has been paying out more and more money abroad for the oil; but it may also be charged again for this oil. When the Arab money which was deposited in the banks and subsequently loaned out has to be written off in the form of bad debts, the U.S. banking system must absorb the loss or pass it on to the U.S. taxpayer, if the government bails out the banks. The present prospects are that many of these loans will never be repaid. How well the U.S. banking system can absorb these losses remains to be seen. Some view this situation as an economic time bomb, which may blow up in the taxpayer's face because of the legal commitments of the U.S. Government to the national banking system.

Thus the transfer of wealth from the industrialized nations (and particularly the United States) is being accomplished in two directions. It goes to the nations which have the oil, such as Saudi Arabia and Kuwait, and to the nations that defaulted on their debts to the United States banks which received and then loaned out the oil money. Thus, the United States, because it has become dependent on foreign oil, is twice a loser in this scenario.

Being self-sufficient in oil does make a difference — a lot of difference. If the United States were still self-sufficient in oil and had not sent billions of dollars abroad to pay for the oil, it would not be in this situation. It was the increasing dependence of the United States on foreign oil which enabled the OPEC countries to raise the price of oil as high as they did and thus produce such a great surplus of money.

No doubt the price of oil would have been raised anyway, but the extent to which it was raised would have been markedly modified if the United States was self-sufficient in oil. This is an example of oil as money, and the problems which accrue to a highly industrialized, high oil-consuming nation which is running out of oil.

Unfortunately, unlike Japan, which long ago knew it had to export to survive, the United States was content with its own large internal market and did not need foreign exchange. But in the past decade, the United States has gone from the world's largest creditor nation to the world's largest debtor nation. This is an almost unbelievable turn of circumstance. Given the balance of payments problem and the ever-rising oil bill, getting itself out of this hole will be a very large task. As oil production in the United States declines, the huge annual oil bill, now nearly $50 billion, will further complicate the balance of payments situation. The U.S. Department of Energy states that by the year 2000 the oil import bill to the United States will reach $100 billion annually, in today's dollars. Domestic oil discoveries will help, for the Alaskan Prudhoe Bay field alone, from only about half its total oil produced, had saved the United States about $135 billion in foreign oil import costs; but unfortunately that field has already begun its inevitable production decline. However, the $135 billion that Prudhoe Bay oil saved the United States in foreign oil purchases was at least that much less money that could have come back to United States banks to be put out in additional problem loans, which, as noted, may ultimately be charged to the U.S. taxpayer.

It is not unreasonable to state that an appreciable part of the Latin American borrowing was wasteful, in that it was primarily used to finance current consumption and budget deficits by governments eager to stay in power through buying off the opposition and sustaining an unrealistic standard of living for its citizens by financing imports. Brazil, however, in all fairness, was one country which, to a considerable degree, did utilize much of its borrowing to build up productive capacity. One of these capabilities was iron mining. Here, in effect, however, the United States financed its own competition, as huge supplies of high grade iron ore are now available from the mineral-rich Brazilian state of Minas Gerias. Some of this ore is shipped to the United States, while, at the same time, the iron range of Minnesota, its higher grade ores depleted and only its lower grade taconite ore left, has difficulty in competing and is depressed. Also, Japan, at one time a buyer of United States ore, now gets some of its ore from Brazil. But even with the projects which Brazil did develop with some of its borrowed money, they are behind in their loan obligations.

In various ways, the world recycling of the huge amounts of oil money which causes these great transfers of wealth is still in progress. When or how it ends no one can say; but the world's money and the world's total economic structure will never be the same again. Oil turned it upside down, and the locations of the world's great oil deposits have had, and will continue to have, profound effects on the economic position of the United States and every citizen therein. The same applies to virtually all other industrialized countries of the world.

It will ultimately be the citizens who will have to pick up all the bills, in various ways — the higher oil costs, and perhaps also the bank problems. Higher taxes, a depreciated currency, and a lower standard of living may be some of these costs. Oil, who has it and who does not, was and remains the principal factor in these problems.

It is striking to consider these profound effects upon the world of a single energy mineral — oil — which 150 years ago was of virtually no concern to anyone at all — one of its chief uses in the Middle East, at the time, being the treatment of camel mange.

As a geologist, I also find it interesting to contemplate that these current events were determined some 40 to 60 million years ago, as the Persian Gulf area slowly sank beneath warm, semi-tropical seas. Untold millions of tons of organic materials were produced here and slowly drifted down to the bottom and were covered by layers of mud which later became rock. Ultimately these events produced what is now the largest accumulation of oil in the world. There is simply nothing else likely to be found to match the Persian Gulf petroleum deposits. The broad framework of our lives is indeed organized and ruled by natural events, over which we had no control and which may have occurred millions of years ago. Continents rise and sink, coal swamps are formed, hot solutions from the Earth's interior put mineral deposits in various places, and the human race must cope with these events of the past.

We like to think of ourselves as being masters of our own destinies; but if there is a predestination in this world, it is in the form of geologic events which laid out the basic geology and mineral and energy mineral geographic framework into which all nations and peoples of the world must fit. It is a mold from which there is no escape.

Dealing with this, in a rational fashion, through the medium of minerals and money will continue to be a major challenge for the indefinite future. Mineral resources, who has them, and their relationship to our industrial societies and our monetary systems will continue to markedly influence the destinies of nations.

Chapter 6

Mineral Microcosms

Mineral and energy mineral resources from many and varied places, are delivered to us, although we may be a long distance from the areas from which these materials come. What happens to one of these areas when the resource is exhausted may be of no concern to us. Other supplies will be obtained. But to the community or the country which produces a given resource from which that community or country derives a substantial portion of its income, the exhaustion of the resource becomes a very serious — sometimes fatal — situation.

One can visualize the ultimate problem of a country or a region with essentially just one resource on which to base its livelihood (and there are some in this situation), by examining a small scale example. The American west is dotted with the names of mining towns which no longer exist. If the community could not find another basis for making a living beyond the original mineral economy on which it was established, it died. A few have survived as tourist attractions, when they are fortuitously located near population centers with more permanent economic bases. Virginia City, Nevada is such an example; founded first on the rich Comstock silver lode, it now hosts curious visitors from nearby Reno. Leadville, Colorado is a similar situation. The citizens in Leadville are now trying to build up the tourist trade, and have tried also to bring in some new businesses. But at an elevation of about two miles, the winters are long and the scene is a bit barren, being just below timberline; and this does not attract a lot of business. Still, some sort of specialty enterprises might wish to locate there. It is a scene different from the ordinary. Being close to an interstate highway helps a bit, but the glory days of Leadville disappeared along with the ore deposits.

The Keweenaw Peninsula of Michigan was the site of the richest copper deposits ever found in the United States, but the only one mine remains. A few small communities have survived but with difficulty; and have only modest economies, of a seasonal nature. Copper Harbor is one of these. Near the tip of the Peninsula, it is picturesque and an interesting place to visit but a difficult scene in which to make a living.

Across Lake Superior, in the iron range country of northeastern Minnesota, the once rich mining towns of Hibbing, Virginia, and Eveleth, and the ore shipping town of Silver Bay have suffered large economic slides. At one time, things were so good in Hibbing, from the economic benefits of the biggest iron ore mining operations in the United States that they had money for all sorts of projects, among which was a "glass schoolhouse" (made of glass bricks). That building and the lavish (for the times) high school with big beautiful chandeliers hanging in the auditorium, along with velvet covered seats, were the envy of the entire State of Minnesota.

But the high grade hematite is gone and iron mining employment is only about a third of what it once was. We fought two World Wars with iron ore from the great Hull-Rust mine, at the north edge of Hibbing. It is an ironic twist to the story of Hibbing that with the reduction in iron mining, some of the residents formed a company to make chopsticks for the Japanese, the first shipment of which was sent in 1987. Post-war, the Japanese have made tremendous inroads on the U.S. steel industry, exporting a lot of cheap steel to our shores; but they get their ore chiefly from Australia and Brazil. Hibbing, once the iron mining capital of the world, still mines some iron but also began making chopsticks for the Japanese. The Japanese said that although they had lost the war militarily, they would defeat us economically; and there are indeed elements of that statement now surfacing. But the cheerful aspect of the Japanese chopstick deal was that chopsticks are used only once; with more than 100 million Japanese, market demand would be steady and perpetual. The chopsticks were made from aspen trees which grow well around Hibbing and represent a renewable natural resource, unlike the depletable iron ore. The factory employed about 15 people. But alas, we had to use the past tense to describe this operation, for it went bankrupt.

Butte, Montana is the site of what is said to have been the "Richest Hill on Earth," according to the sign at the city limits. It may have been once, but the copper ore is largely depleted, although from time to time efforts are made to revive the main mine, or bring other nearby smaller mines back into production. The main mine, the famous Berkeley Pit, is now a big hole on the north edge of town, partly filled with water. The smelter at Anaconda, a few miles to the west, stands abandoned. It is doubtful that it will ever smelt another pound of copper.

Just a short distance west of Ely, Nevada lies a deep and colorful hole, the copper pit at Ruth. But the one-half of one percent copper ore could not justify removing the increasing amount of overburden, as the mine had to go deeper to find the ore. This operation is now abandoned, as is the smelter which went with this mine, located at the now essentially deserted community of McGill, across the valley east of Ely. Ely is a nice town and it survives, after a fashion, as a supply center for ranching activities in that part of Nevada. Probably the recently created Great Basin National Park, some distance to the east, near the Utah border will bring in some business. But the motels which have been built are well beyond what the community needs now; and if you visit Ely, you will have your choice of nice rooms, at very competitive prices. And even the business which has had a long and successful history around the world, traditionally marked in the United States with a red light, has been substantially reduced. The mining clientele is gone.

The same general picture can be seen in the Coeur d'Alene mining district of northern Idaho. Once the site of the largest zinc mine in the United States and a huge smelter to go with it, the Bunker Hill, the valley now survives largely by tourism, some modest mining, some logging, and a little ranching. It is an interesting place to visit, if one wishes to see numerous waste dumps of mines, now long abandoned, and see how ingenious people built houses precariously perched on the hillsides of the steep gulches of the area. Perhaps the mines will some day come back — some of them. A few still operate from time to time. The largest silver mine in the United States, the Sunshine, opens and closes with the rise and fall of the silver market; but the boom days are quite likely gone forever.

Ajo, Arizona is a place where you can buy a house very reasonably. It was chiefly a company town, and Phelps Dodge operated the large copper mine there. But it, too, has seen its day; and that day has departed.

Probably the newest and most striking ghost town of the west is Parachute, Colorado, or, more specifically, the housing complex built across the Colorado River from Parachute, in anticipation of a big oil shale boom. This community became a ghost town before much of it was even occupied. There are, in fact, a few people living there, but many of the housing units never have been lived in.

Forty years ago, as I was driving west to do geological studies in Nevada, I passed a sign in western Colorado which read "Oil shale boom. Get in on the ground floor." You can still get in on the ground floor, but it may prove to be the sub-basement, as it has been up to the present. With the great rise in oil prices in the 1970's, some people thought that oil shale (from which you get shale oil) was to finally become big business. Several major oil companies already held large lease positions in the area. Other companies bought in, consortia were formed — the most notable of which was Colony

Development Corporation. Big plans were made to operate up Parachute Creek, where the richest oil shale deposits are located.

A sleepy little community named Grand Valley (subsequently renamed Parachute, in deference to the importance of adjacent Parachute Creek oil shale deposits) began to expand. On the plateau across the Colorado River, south of Parachute, a shopping area was planned and partially completed. Numbers of housing units were built and several large apartment complexes of a grand nature were constructed (Big oil companies do things "right."). A good-sized motel sprang up (at least part way) on the north side of the highway, right at Parachute. Then came the bad news in April of 1982. The $5 billion Colony oil shale project, the largest in the nation, was being shut down. The auxiliary facilities were in place; but the shale oil plant, which was to keep it all going, never got built — a really classic example of the expression "putting the cart before the horse."

So Parachute now is in marked decline and the oil company is trying to market its very nice housing project as a retirement village. Nicely landscaped and well built living units can be purchased very reasonably. If you want to retire and "get away from it all," this is the place.

Thus, as quoted elsewhere in this volume, the saying which has been around oil shale country for so long — "shale oil is the fuel of the future, and always will be" has again come back to haunt the region. At some future time, Parachute will probably come back to life as oil really becomes scarce and expensive; but, for the moment, Parachute is again a quiet community, although Unocal still operates a modest shale oil plant. But it may not survive much longer.

Down the road to the west of Parachute is Grand Junction, Colorado. During the uranium boom of the 1950's, Grand Junction was the center of great exploration activity in the Colorado Plateau. But with interest in uranium now at relatively low ebb, Grand Junction, like Parachute (but not quite so severely) has suffered a substantial decline. It is now back to being chiefly a community to supply the area ranchers, and a place to stop and buy gasoline when you drive from Denver to Salt Lake. If you want to stay overnight in Grand Junction, you will find an abundant supply of motels, with reasonable rates.

Oil generally tends to be a longer producing resource than are the many small mineral deposits on which the now-defunct western mining towns were built. But oil, too, is finite; and the question increasingly being asked by many smaller communities and even some larger areas is "what do you do when the well runs dry?" Such is the question being asked in Van, Texas which, starting about 1930, when oil was first discovered under the town, has been financing more than three-quarters of the city expenses from oil revenues derived from about 370 wells, pumping at various places in the community, including several in the yard outside the Van High School.

But the edge wells of the Van Field are now drying up (more nearly correct — going to water), and the revenues are shrinking and will ultimately stop. Who or what pays the approximately 80 percent of city costs then?

In contrast to Van, Giddings, Texas was an example of a new oil boom town. In 1976, only five wells were drilled in the vicinity, whereas in 1981, 945 were drilled. The housing shortage in Giddings was so acute as workers flocked in, that one enterprising citizen tried to open a motel using old oil storage tanks as bedrooms. The Giddings population abruptly rose from 3,900 to more than 8,000. More than 110 ribbon cuttings marked the opening of new businesses. In June of 1980, an "Oil Appreciation Week" was held, with the theme "Praise God from Whom all oil doth flow." The police force was increased from 4 to 17, and school taxes were cut by more than 50 percent.

An acre of farmland at the edge of town which may have produced about $200 worth of peanuts per year might bring a thousand dollars for a drilling lease. Thirty new millionaires emerged as a result. So, while Van, Texas watched its oil production decline and the town slump accordingly, Giddings, Texas boomed — for a time, that is. The oil price drop, which began in 1982, pricked the Giddings oil balloon. Workers were laid off, builders found themselves suddenly with new houses but no market. And so these mineral microcosms had their day. First Van, Texas, and then Giddings.

By definition, microcosms are small and they also tend to have a rather ephemeral existence. A local small but rich mineral deposit is the basis for the boom and also sets the stage for the subsequent "bust." Without the brief mineral bonanza, such things would not happen; but for a short time, the mineral does shape the destiny of the community. Sometimes it has been the sole basis for the existence of a community, as the many ghost towns of the west can testify.

From here we shall go on to stretch the definition of a microcosm to include a territory as large as a state, and look at

Louisiana. Huey Long, the great Populist, built his career by publicly battling the oil companies in the name of the common man; but in 1934, he secretly threw in with them, when he and several others formed a company called Win or Lose Corporation. They bought state mineral leases and resold them, at great profit, to out-of-state oil companies. One of these was a small Texas-based drilling firm called the Texas Company, which would eventually grow into the giant Texaco.

For more than 50 years, the State of Louisiana has been a very highly rewarding oil exploration area, with drilling success rates far beyond most other regions. During this time, about 15 billion barrels of oil and 120 trillion cubic feet of natural gas have been produced. On each barrel of oil and on each one thousand cubic feet of natural gas the State of Louisiana has levied a tax. And the money rolled in — in huge amounts. The state government of Louisiana for many years obtained about 40 percent of its income from petroleum, and the legislature had so much money it didn't know what to do with it. Salaries were raised. All sorts of social programs were started, including a denture program for senior citizens, which came to be known as "the right to bite." The first $75,000 value in the homestead was exempted from taxes, which resulted in 85 percent of the citizens currently paying no property taxes.

To get at the oil in the marshlands, draglines were brought in to cut channels and provide access for the huge drilling equipment. This changed the ecology of these areas. What were once prime fishing and trapping areas became biological wastelands. But now oil and gas production is in marked decline. One state report estimates that by the year 2000, Louisiana will be 97 percent depleted of oil and 90 percent depleted of gas. Such estimates can be wrong in detail; but the trend is clearly there. Louisiana is now experiencing oil and gas withdrawal pains; and this pain will persist, for some time to come. Unfortunately, also, in trading for the quick riches of petroleum, many of the once highly productive marshlands, which produced renewable natural resources, such as fish, furs, and shellfish, have been badly damaged. The Cajuns (the French of southwest Louisiana) who went into the oil industry as workers (from being fishermen and trappers) now find that the oil business is declining but they cannot go back to trapping and fishing as they once did. Oil has changed their lives and the economies of the communities for a very long time to come — perhaps, in a practical sense, forever.

Also, during the oil boom, Louisiana spent money on many capital improvements, such as bridges, buildings, and roads. But these subsequently need to be maintained. Now that the oil revenues are declining, the money to maintain this infrastructure is not there. And from petroleum, at least, it never will be again. Louisiana does have a good agricultural base, and has a substantial petrochemical industry; but eventually more and more of the raw materials for the petrochemical industry will have to be imported. Furthermore, and perhaps more important than the probable increase in the cost of obtaining petrochemical raw materials, is the fact that as countries which produce petroleum try to upgrade their product before shipping it out as merely raw materials, as in the past, the need to build petrochemical plants of their own becomes apparent; and these, in turn, can effectively compete with those of Louisiana. Saudi Arabia has built just such an extensive petrochemical complex, and continues to expand it. Kuwait is doing the same.

The rapid exploitation of petroleum resources in the United States is without precedent in the world; and Louisiana is a good example of the rags-to-riches story which accompanied this event. But now the riches, which are non-renewable, are gradually diminishing; and in fact, if the once highly productive marshlands of that State have suffered long-term (perhaps in some areas, irreparable) damage from oil operations therein, it may be that Louisiana will ultimately be poorer than before the oil industry arrived. Everything has its price. It was a fine ride while it lasted.

As a result of this change in economics, there is also a rather sad peripheral result becoming evident. In Morgan City, Louisiana, once a booming area for building and maintaining offshore drilling rigs, some authorities have noted an increase in social ills, because of the economic pressures resulting from the decline in our petroleum industry. One might suggest that this situation could also develop in larger areas, even countries, as a once rich mineral resource declines against a still-rising population, with rising expectations. The results could be explosive. To some extent this happened in Algeria in 1988, when declining natural gas revenues caused an austerity program which resulted in riots, and the death of more than 400 people.

Like Louisiana, the great State of Texas, premier oil-producing state of the United States, also has seen the peak of its oil industry. Texas oil reserves in 1960 were nearly 15 billion barrels. By 1984, the oil reserves were estimated to be about 7.6 billion barrels and still declining. Texas, too, is suffering from petroleum withdrawal pains — as the petroleum is withdrawn, the pain progresses. Neighboring Oklahoma, like Texas, is suffering a similar situation, as are New Mexico, Colorado, and Alaska.

Houston, which had never really seen a depression before, with the oil price crash of the 1980s saw people simply moving from houses, giving them to the banks and savings and loan associations. This became true all through the oil states, including Louisiana, Texas, Oklahoma, Colorado, and Alaska, and accounted for a substantial part of the savings and loan financial debacle, which is still with us as of this date, and may eventually cost the American taxpayer in excess of 200 billion dollars, as he/she is called upon to rescue the Federal government's deposit insurance fund guarantee.

The Alaskan petro-cousin in the bust part of the oil cycle saw Anchorage alone lose about 12 percent of its residents and one in twelve persons in the state with mortgages on their homes lost them. In all of Alaska, some two and a half million square feet of store and office space became vacant. The Alaskan oil industry, through all its ramifications, produces 89 percent of the gross product of the state; and state government gets 85 percent of its money from oil royalties and taxes. State employment and services were cut drastically with the oil bust; yet one vestige of the affluent days remains — the payment of some of the state oil income yearly, to every Alaskan citizen. In 1982, this program was voted in and began that year with $1000. This annual payment continues, in spite of the disastrous condition of the state treasury. By this law, in 1988 the state had to give out some $423 million. Once on welfare, it is hard to wean the citizens off. And this is proving true also in some of the oil-rich countries, which started large welfare programs.

Since the discovery of oil there, the State of Alaska has received more than $25 billion in taxes and royalties from that source. Most of this comes from the giant Prudhoe Bay field on the North Slope. But this field is in decline, portending an attendant drop in state revenues. There is an estimated 3.2 billion barrels of recoverable oil in the region east of Prudhoe Bay, which includes the Arctic National Wildlife Refuge. Development of this area for oil would help Alaska's budget but the idea has split Alaskans over the possible environmental effects. With 85 percent of Alaska's state income coming from oil, the dispute is sure to be a large one. But then, of course, one may wonder what ultimately happens when there is no more oil to produce income. And the day will come.

In the "oil patch" states, public budgets are being revised; sales, liquor, cigarette, and gasoline taxes and other sources of revenue are being raised. The economies must shift to new, more sustainable and renewable bases and away from the one-crop mineral resource revenues. It is to be hoped that such can be found but the economies will be forever changed.

The discovery, development, and decline in the production of petroleum and other mineral resources has a marked effect upon communities and states where these resources exist. But there is also a wider ripple effect through the economy in areas perhaps far from the site of the resource. A great many microcosms (individual enterprises, or projects, or groups of people) which may be far removed from the oil wells or mines can be affected. When the oil slump hit in the mid 1980s, Floating Point Systems of Beaverton, located in oil-less Oregon, but maker of high-powered computers widely used in the oil industry, saw its orders cut drastically, and layoffs resulted. The nationwide Public Broadcasting System (PBS) of the United States saw the sponsorship of its programs by oil companies markedly reduced; and oil companies had been a significant part of the PBS budget, making up more than half of that network's million-dollar or more annual contributors. The helicopter business is markedly affected by conditions in the oil industry, as at one time more than half the civilian helicopters in the United States were sold to ferry workers and supplies to offshore drilling rigs. By recent count, more than 200 of the 800 helicopters in the Gulf of Mexico area were surplus.

The Girls Clubs of America, Inc. felt the decline of the oil industry as ARCO and Exxon cut their donations by more than half. Even the TV evangelists felt the effects of America's declining position in the world oil industry, as an appreciable part of their income tends to come from the so-called "Bible-belt" of the South and Southwest, which is also the "oil patch" that is Louisiana, Arkansas, Oklahoma, and Texas. Some of them, in making the case for more generous denotations, emphasized in their broadcasts that the oil slump was causing a substantial reduction in the financial support of their ministries.

Thus in obvious ways, and in more subtle ways, mineral and energy mineral resource exploitation has a large, and in some cases, an overwhelming impact on many small units of population and some not so small, such as regions and states. In some local cases where the depletion of the resource is essentially total, a community may simply disappear. Many have, and more will. But on the other hand, others from new discoveries or the application of new technologies to waning resources, will spring up overnight, or stage a come-back for a time.

All of these mineral microcosms just described are local or regional in nature; but perhaps they illustrate in small and relatively simplified form the problems which certain nations may face, if they are dependent largely on one or a few depleting resources. Some of these countries are discussed in the next chapter.

Chapter 7

The Chiefly One Resource Countries

The designation of a "chiefly one resource country" as used here means a country which does not have a significant industrial base and therefore must depend, to a considerable extent, on the export of a particular mineral resource in order to obtain foreign exchange, by which to buy manufactured goods which characterize the modern world. Some countries with very limited mineral resources, such as Japan, do very well in terms of foreign exchange, by virtue of having a successful, competitive industrial complex, which can export manufactured goods to pay, in turn, for the import of raw materials. Switzerland is another such example. But many countries do not have such a base, particularly the smaller countries.

Also, to have a variety of minerals, there must be a diversity of geology, for different minerals are found in particular geologic settings; platinum, tungsten, and oil, for example, occur in very different natural situations. Accordingly, the smaller the country, the less likely it is to have a wide spectrum of mineral resources. Even some larger countries, given the random nature of the distribution of minerals, may be essentially one resource nations. Such is the case of Saudi Arabia.

Here we examine some representative countries which obtain more than half of their foreign exchange from a single mineral. A second group of countries is also considered, which obtains a substantial part of its foreign exchange from a single mineral, or group of minerals.

Algeria. This North African country is large with an area of about 919,000 square miles and a population of only 20 million; but the economics of the area are difficult, as evidenced by the widespread rioting of the civilian population in late 1988. Only about three percent of the land can produce crops; the rest is inhospitable desert and mountains.

The principal resource which Algeria has, by which to obtain foreign exchange, is petroleum, to some extent upgraded to refined products. Petroleum, in one form or another, constitutes 92 percent of its exports. A small amount of iron and steel make up the rest of the outgoing foreign trade. Without petroleum, Algeria would have very little with which to buy the world's goods, and because it has a very limited manufacturing base, most items of modern day living, including such vital items as medical supplies, have to be imported. The riots, already alluded to, and in which more than four hundred people were killed by the army, were caused by food shortages and rising prices, brought about by the drop in petroleum prices in late 1988, which substantially cut the Algerian income. With its population still rising, and with petroleum, Algeria's main source of foreign exchange, one might wonder what is going to happen in that country when the petroleum deposits are exhausted, as will eventually happen.

Angola. This southwest African country has been torn by internal strife for a number of years. This has interfered with the shipment of iron ore, which exists in large quantities. However, foreign oil companies have kept the flow of oil going, for the oil is largely from offshore operations, whereas the iron ore comes from the interior of the country, made largely inaccessible for ore transportation because of the civil war. Being offshore, oil production has not been disrupted. Petroleum and related upgraded petroleum products make up about 85 percent of Angola's exports. Diamonds represent another 10 percent of its export value, so, combined, these two minerals make up about 95 percent of Angola's ability to generate foreign exchange. Again, like most small countries, there is no significant or broad manufacturing base. Thus most items, beyond those representing a subsistence level of living, have to be imported. Petroleum pays almost all of that bill.

Bahrain. This is a group of islands in the Persian Gulf, with a total area of about 225 square miles. The islands are low, with the highest point less than 500 feet above sea level. Oil was discovered in 1931; and since then, a large refinery complex has been built, which also refines oil from other areas of the Gulf.

Government revenue was derived almost entirely from petroleum until recently when, with the construction of a causeway from Saudi Arabia to Bahrain, various worldly enterprises have brought in some money. Although Bahrain is nominally Moslem, the strict rules of that religion are not rigorously enforced in Bahrain. The result is that gambling and drinking and other aspects of western world living have come upon the scene. Moslems from oil-rich but other less-liberal areas visit easily via the causeway to Bahrain, to sample western "culture." Thus Bahrain's economy is supported in two ways by oil: the oil within the country, and the oil wealth which comes from neighboring countries via the causeway. Without oil, Bahrain would revert to what it was — a fishing and pearl diving archipelago, with humidity which is almost always above 90 percent. The pearl diving and fishing have been greatly reduced at present, because of oil income; but the humidity remains. However, oil and gas come to the rescue again, fueling generating plants producing electricity to run air-conditioners, which make Bahrain much more livable than previously.

Brunei. This tiny country, on the northwest coast of Borneo, with an area of only about 2200 square miles and 200,000 people, has one of the highest per capita incomes in the world. Oil is the source of the wealth. Brunei's economy depends almost entirely on petroleum, which employs seven percent of the working population but accounts for more than 93 percent of its exports. Petroleum is almost the entire basis for the Brunei economy. Without petroleum, what will Brunei be like? Can a now more numerous and much more affluent people make an orderly transition back to a more austere, non-oil economy?

Ecuador. This delightful and very scenic country is one of the smallest of the Latin American republics in South America. It might (quite honestly and without prejudice) be called a "banana republic," for it is the largest exporter of bananas in the western hemisphere. But the chief source of Ecuador's hard currency; that is, foreign exchange, which is widely accepted internationally, is oil. Oil provides about 70 percent of what Ecuador earns abroad. Being such a small country, Ecuador has a limited manufacturing spectrum, with the result that a great many things have to be imported.

Although Ecuador belongs to OPEC, it is an unstable member in the sense that it tends to produce whatever oil it can, ignoring any quota OPEC may have decreed; and sells the oil largely on the spot market, for whatever the current traffic will bear. Sometimes it is below OPEC's benchmark price, but OPEC is inclined to overlook the infraction, as Ecuador's export of oil is usually less than 150,000 barrels a day. Yet this brings in nearly three-fourths of Ecuador's foreign exchange. Can bananas forge ahead and, in the future, make up the difference when the oil is gone? Now, even with the oil income, Ecuador has accumulated more than $8 billion in foreign debt; and oil income is the chief source Ecuador's ability to repay the debt. When an oil pipeline was severed, because of a huge landslide, Ecuador had to suspend payments on the international debt until the pipeline could be repaired and oil again flowed to service the debt. On a per capita basis, each Ecuadorian man, woman, and child has a foreign debt of nearly $1,000. If oil production falls, there is little likelihood that the debt can be either serviced or repaid. Eight billion dollars is a lot of bananas!

Indonesia. This is the fifth largest country in the world, in terms of population. It is a group of islands including Sumatra, Java, Bali, part of Borneo, and the western half of New Guinea.

There are more than 3,000 islands, with a land area of about a half a million square miles. Prior to World War II, it was a Dutch possession, called the Dutch East Indies. The Japanese invaded these territories immediately after Pearl Harbor, to obtain their oil supplies. After the war, it became an independent nation. The Dutch never returned. Indonesia is the largest oil producer in the Far East and is the tenth largest producer in OPEC. Oil, until recently, has constituted about two-thirds of its export values. Recognizing this excessive dependency on one resource, the government has been working to strengthen the competitive position of its other exports in world markets; and recently, petroleum sales have constituted only about half of the total value of exports. However, government revenues remain heavily dependent on petroleum, and the economy, as a whole, remains very sensitive to petroleum markets.

Iran. This Persian Gulf nation, much in the news during the 80s, because of its conflict with Iraq and the disruption of oil shipping lanes, is a Muslim but not an Arab nation. The huge oil deposits of the Persian Gulf area are sometimes called, by the predominantly Arab countries of the region, a "soft loan from Allah." Apparently, however, in the case of Iran, being Muslim is sufficient to get in on the largess from Allah, and one does not have to be Arab also.

Before the outbreak of hostilities between Iran and Iraq, Iran obtained about 94 percent of its export income from the sale of oil and its derivatives. This diminished during the war, as Iraq made concentrated efforts to destroy some of Iran's shipping facilities. Iran, in turn, made a great effort to keep their oil exports flowing, as they were a major source of income to pay for both military and civilian supplies; even so these shipments were reduced somewhat during the conflict.

Iraq. This country on the northwestern end of the Persian Gulf is both Muslim and Arab. It has been, for a number of years, almost totally dependent on oil for its export revenues. Its oil income was severely impaired for a time by the war with Iran; but eventually pipelines built to reduce dependency on the Persian Gulf as an outlet for the oil, brought Iraq's oil output almost back to pre-war levels. Before the war, Iraq received 99 percent of its foreign exchange from the sale of oil.

Jamaica. This Caribbean country has substantial aluminum ore (bauxite) deposits, and has been one of the world's ranking producers. However, the ore deposits are beginning to be depleted. Nonetheless, bauxite and a slightly upgraded product, called alumina, constitute more than 60 percent of Jamaica's export values. Jamaica has few basic industries and, like other small island economies (Its area is only about 4,400 square miles), has to import virtually everything except for some agricultural products. Income from bauxite and its alumina derivative provides most of the money used to buy manufactured goods, and even some food supplies.

Kuwait. If one wishes to compare the value of mineral resources per square mile, of a territory or a whole country, Kuwait is almost in a class by itself. Lying at the northwestern end of the Persian Gulf between Iraq and Saudi Arabia, this small country, with an area of approximately 7,000 square miles, and a population of about 1 1/2 million people, has beneath its soil more than three times the total oil reserves of the United States, in its 3 1/2 million square miles. The production from Kuwait's approximately 92 billion barrels of oil reserves give it one of the highest annual per-capita-gross-national-product figures in the world — more than $30,000. The role which Kuwait plays in the world economy (because of its singularly large oil reserves) is greatly disproportionate to the size of the country. Management of the vast amounts of money which have poured into the country, as a result of its oil wealth, has at times created some chaotic financial situations. But Kuwait, as noted elsewhere in this book, is trying hard to make well-placed investments abroad, so when their oil runs out, the oil legacy will survive. Kuwait is keenly aware that it is a one-crop-one-resource country; and good investment decisions are crucial to its future.

Liberia. This west African nation was established by several American philanthropic societies, to provide a place where freed American slaves could return to Africa. The state was formally established in 1847. The imprint of America remains on the country in several ways, one being that the unit of currency is called the dollar.

Unknown at the time it became an independent nation was the fact that it possessed some exceedingly rich iron ore deposits, conveniently located near the coast. The result is that, at the present time, Liberia can lay down high quality iron ore on the East coast of the United States, competitively priced with ore coming from the Mesabi Range of Minnesota. Iron ore concentrates are currently (by far) the most important of Liberia's exports. Gold and diamonds, in small amounts, have also been discovered and are exported. Liberia's foreign exchange is therefore derived principally from minerals — mostly iron.

Libya. Mostly desert, in part mountainous, this country has an area of about 680,000 square miles, or approximately one-fifth the size of the United States; but it only has a population of four million. Moreover, only a very narrow strip along the coast can be successfully farmed. A marginal grassland grazing area lies immediately to the south; but this thins out rapidly into the Sahara Desert and, at best, the grass is sparse. Until oil was discovered, Libya was a country of no great importance, with virtually no impact on the rest of the world. But the discovery of oil changed that. Oil rapidly became the largest source of government revenues, and allowed Libya, in various ways, to buy into the twentieth century. Without oil, Libya would still be what it had been for centuries, a nation of small farms and herding operations. Libya is markedly a one resource country, with oil consistently making up from 90 to nearly 99 percent of total exports. When the oil is gone, the economy will have to be greatly restructured toward a simpler life. Change is inevitable, and there is nothing in sight to replace the oil.

Mexico. This country was an oil producer early in this century; but substantial oil production was not developed until fairly recently, the big oil boom starting in the 1970s. Mexico now produces about 2 1/2 million barrels of oil a day; and appears to have oil reserves at least as large, and probably much larger (some estimates say two or three times) than those of the United States. This somewhat imprecise statement about Mexican oil reserves stems from the fact that as the Mexican oil boom got underway, there was an economic (and political) need to make the situation look as good as possible, because it was largely on the basis of oil reserves that foreign banks were willing to loan money to Mexico. Some of the banks apparently believed what now seem rather inflated oil reserve figures, and loaned money accordingly to their later regret. The most reasonable figure seems to be about 48 billion barrels (compared to the USA, with about 25 billion). In any event, petroleum and its products make up (depending on world prices) between 60 and 75 percent of Mexico's export value. Most of this goes to the United States.

Nauru. This is the world's smallest independent nation. It is an island which lies almost in the center of the Pacific Ocean, and has an area, not figured in square miles, but in acres — 5,236 of them. And these acres are composed primarily of superphosphate, which is easily processed into a very rich agricultural fertilizer. Nauru's export trade consists entirely of this phosphate, which goes to Japan, New Zealand, and Australia. Much of Nauru's land is devoted to phosphate mining and after the mining is done in a given area, what remains are gray, inhospitable pinnacles of rock. For the approximately 4,000 Nauru citizens there are only about 1,100 acres (less than two square miles) available on which to live. This is the world's best example of a nation dependent upon a single mineral resource. Unfortunately, it will be gone in only a very few years; the citizens of Nauru know it and what they are doing about it is described in the next chapter.

New Caledonia. Like Nauru, this is an island in the Pacific. It is an independent country but is more or less under the protective wing of its earlier French possessors. It has what is probably the second largest nickel deposit in the world. Nickel is regarded as a strategic metal, useful to industry in many ways, including the production of war materials. New Caledonia's foreign exchange is obtained almost entirely from nickel. It is the only substantial resource that island nation has. As nickel goes, so goes the New Caledonian economy. Being a small island, it has to import nearly everything, except some locally grown foodstuffs. But, nickel pays the bills at present. In the future, when the nickel is gone, there is no obvious replacement.

Niger. Several countries in Africa are substantial uranium producers, or have the potential of being so. These include Namibia, South Africa, Gabon, and Niger. Of these, Niger is the most vulnerable to the uranium market; and when the uranium boom ended in the early 1980s, Niger was dependent on uranium for 84 percent of its foreign earnings. Since that time, the uranium market slump has markedly depressed the Niger economy. There are few other resources, even in total, to take the place of the atomic metal as an income for the country.

Nigeria. This is the most populous African nation with nearly 100 million inhabitants. Nigeria's dependence on oil for its export earnings is almost total, making up, at various times, between 80 and 95 percent. For some other resource, which would probably have to be an agricultural one, to take up the slack, when the oil is gone, seems an impossibility. Because of Nigeria's great dependence on oil to lubricate its industrial, as well as its social and political systems, Nigeria is one of the countries of OPEC which tends to frequently cheat on its production quota.

Oman. More properly called the Sultanate of Oman, this country, with an area of about 105,000 square miles, has a coastline about 1,000 miles long, on the southeast end of the Arabian Peninsula. It was a country of date palms and camels until 1964, when oil was discovered in commercial quantities. Now oil provides essentially all government revenue. Copper was subsequently discovered and a small copper mining and refining industry (using the natural gas from the oil fields) has developed. In total, petroleum, with a little help from copper, provides more than 90 percent of government monies and foreign exchange.

Qatar. Probably not one person in a hundred can tell you the location of this country, but it has a per capita gross national product more than twice as large as that of the United States. Qatar occupies an area of about 4,000 square miles (the whole of the Qatar Peninsula) which juts out into the southern part of the Persian Gulf, adjacent to Saudi Arabia. It is a region of sand, gravel, and some limestone ridges, and was, until very recently, an exceedingly poor country, dependent chiefly on fishing, pearl diving, and some trading. It was a subsistence existence at best. Now it has a world-class per capita income. Petroleum did it. Qatar sits atop what appears to be 10 percent or more of the total world gas reserves, in what is called the Northwest Dome Field. To use these gas deposits, petrochemical and fertilizer plants have been built; but it is all petroleum based. There are virtually no other resources, except bare subsistence agriculture and a little fishing.

Saudi Arabia. Holding the world's largest oil reserves beneath its sandy soil and below the shallow waters of the adjacent Persian Gulf (Arabian Gulf to the Saudis) Saudi Arabia derives nearly 100 percent of its income from petroleum. Before the discovery of oil, Saudi Arabia was an obscure nation of a few million people, many of them nomadic and dependent on a grazing economy, small farms, and a few fishing and pearl diving villages. Oil alone made Saudi Arabia a world economic power, and oil remains virtually the only substantial resource in that country. Saudi Arabia and its oil are discussed in several other places in this book, including a special chapter, *The Luck of the USA and Saudi Arabia.*

Suriname. Prior to 1975, called Dutch Guiana, located on the north coast of South America, Suriname is an area of heavy rainfall and a warm climate, which, in the course of geologic time, have combined to weather and leach some of the rocks to form a rich residual deposit of bauxite, the chief ore of aluminum. Today, bauxite and its concentrate, alumina, along with some aluminum metal from one refinery, constitute about 70 percent of Suriname's total export value. Most of the aluminum, in its various forms, goes to the United States.

Trinidad. This island, lying a few miles northeast of Venezuela, was initially famous for the presence of the Pitch Lake, an area where tar oozes from the ground and can be mined. Many of the roads of England (of which Trinidad was once a colony) and of western Europe were initially paved with asphalt from the Pitch Lake of Trinidad. Later, oil was discovered both onshore and in near-offshore areas. Based on this, Trinidad developed ammonia and fertilizer plants, petrochemical plants, and cement plants (The chief cost of cement is energy, for burning limestone, and Trinidad uses natural gas for that, because the gas which is produced along with the oil cannot conveniently be exported.). All this is based on the presence of petroleum. The importance of oil to Trinidad is such that its currency carries an engraving of an oil derrick, and well it might, for oil is, indeed, money to Trinidad, with 90 percent of its export income derived from that source, in one form or another. More recently, however, oil production and oil-related income has been declining and there are no other apparent resources available to make up the difference.

The result has been close to disastrous. Suffering from both a declining oil production and the collapse of oil prices, in the mid-1980s, the gross national product dropped from $4.45 billion in 1982 to $2.82 billion in 1987 and continues to drop. Per capita income fell from $7,060 to $3,380 during that same period. Government revenues, 70 per cent dependent on oil, dropped from $1.67 billion in 1982 to less than $1.18 billion in 1988. Unemployment doubled to 22 percent.

Meanwhile, Trinidad's population continues to increase rather rapidly. Harder times have arrived and more may lie ahead. It is an excellent example of the problems confronting a largely one-resource country.

United Arab Emirates. This is a confederation of several sheikdoms, the most important of which is Abu Dhabi. It lies on the south end of the Persian Gulf and has an area of about 32,000 square miles, and a population of slightly over a million people. Like nearby Qatar, it was a relatively poor area until oil was discovered, principally in Abu Dhabi. Proven reserves in these 32,000 square miles are now estimated to be more than 90 billion barrels, which are almost four times those of the United States. Oil here, as in all the oil-producing nations bordering the Persian Gulf, is by far the most important export and source of revenue.

However, the coming of oil was not an unmixed blessing to the United Arab Emirates, for the prospect of this wealth brought in thousands of poverty-stricken people from adjacent territories. They greatly taxed the sanitary and housing facilities of the UAE and also brought with them diseases which had not been previously known in the United Arab Emirates. But for the moment at least, oil has made this Emirate confederation wealthy beyond what anyone there could have dreamed, a few decades ago.

The petroleum age here, as in the case of most other countries around the Persian Gulf, will be remembered as the golden age. The question, however, is what lies beyond oil? Can some sort of infrastructure and technology or commercial enterprise be put in place, to be permanently productive of something of value, or, in some other way, a source of income beyond the time when oil ceases to do the job? Petrochemical complexes now being built are excellent sources of current income but will not do the task for the indefinite future, for they, again, are based on the finite resource of petroleum. In recognition of that fact, at Dubai, the principal city of the UAE, a free trade zone is being built, so as to establish this area as a trading center. The UAE's approach to the problem of "beyond oil," Kuwait's approach, and how other countries are handling the matters of ultimate mineral resource depletion are taken up in the chapter *Mineral Riches And How They Are Spent.*

Venezuela. Petroleum brought Venezuela into the twentieth century, largely through the great oil discoveries in the Lake Maracaibo region of northwestern Venezuela. An affiliate of what is now Exxon Corporation, Creole Petroleum, was the principal developer; but now all oil operations in Venezuela have been nationalized. Venezuela has other mineral resources, chiefly iron ore (a mountain of it, called Cerro Bolivar), and some reasonably good bauxite deposits. But for all these other resources, oil, marketed through the state-owned company, Petroleos de Venezuela, contributes 90 percent or more of that nation's foreign exchange income. It is still largely a one resource country, in terms of trading with the rest of the world.

Zaire. Once called the Belgian Congo, Zaire became independent in 1960. Located on part of the great African mineral belt, Zaire obtains most of its foreign exchange from the mining of copper. Zinc and cobalt also are important, but copper is by far the largest export item. Without copper, and the much lesser amounts of cobalt and zinc, coffee (now far down the line in total value) would be the principal export; but the income from it would be small, compared with that earned currently by copper. Copper is the coin which allows Zaire to buy at least part way into the modern world.

Zambia. Next door to Zaire is Zambia, also astride the rich mineral deposits of south-central Africa, and also, like Zaire, a major copper producer. That metal accounts for nearly 90 percent of Zambia's export value.

We have been considering some of the nations in which more than half of their foreign exchange is currently derived from one mineral resource. There are a number of other countries where a

97

given mineral resource is important but less than half of their exports, in terms of value. In some cases, these countries also export other minerals, so in total, it is minerals which do carry the foreign exchange load. In other countries, there are agricultural products and other items which also help to obtain foreign exchange, and the minerals which are exported make up less than half of the total but still are of a significant amount. We here consider some of these:

Australia. This country (which is actually a whole continent by itself) is commonly thought of as exporting chiefly wheat, beef, lamb, and wool. But actually, the principal export in value is coal. Australia is also now the world's largest producer of bauxite, after discovery of tremendous deposits in the northwestern part of the country.

Bolivia. Mining is the most important single industry in Bolivia with tin making up nearly half the value of production. Minerals (in total) make up about half of Bolivia's exports. Bolivia does not have many manufacturing plants and, as a result, must import most products. Without minerals, this ability would be cut in half.

Chile. Copper, for Chile, accounts for about 48 percent of its exports; and when other minerals are included, the total is more than one half of Chile's exports. Copper is likely to continue to bulk large in Chile's economy, perhaps even more in the future than at present, for Chile holds an estimated one-fifth of the world's copper reserves.

Egypt. This is the largest of the Arab countries, in terms of population — now about 50 million. It is also among the poorest. Imports in recent years have been more than twice the value of exports; and Egypt is kept afloat, to a considerable extent, by the generosity of its other oil-rich neighbors. Egypt itself does have some oil production, about 800,000 barrels a day, and crude oil and some upgraded petroleum products make up about 42 percent of Egypt's export values. This is more than the value of cotton, for which Egypt has long been famous. Cotton is only about 34 percent of its total exports. Egypt probably could use all the oil it produces, internally, to replace the more primitive forms of energy from animals and humans. But with a large population, there are many hands to do the work; and oil is urgently needed to provide foreign exchange.

Morocco. This interesting and colorful country has a good agricultural base, and, situated near Europe and with a mild climate, it sells most of its agricultural produce to Europe, especially in the winter. But Morocco also has a unique position among world mineral producers, for it has the world's largest phosphate deposits. Every living cell must have phosphorus, and phosphate resources are very unevenly distributed in the world. Morocco's phosphate and its derivative, phosphoric acid, make up about 40 percent of Morocco's exports. And because phosphate is so important for life itself, Morocco is fortunately situated on a very large deposit of this important mineral. Good geological luck for Morocco!

Namibia. This country, previously known as Southwest Africa, gets 40 percent of government revenues from royalties due from diamonds recovered on the coast. Namibia also has what is currently the world's largest uranium mine. Between diamonds and uranium, Namibia is substantially more than half dependent for revenue on these two minerals.

Norway. This delightful and very scenic country unfortunately has only four percent of its land that is arable. One way or another, Norway has had to turn to the sea, first by fishing; more recently it has found another reason to go to the sea, by virtue of the large oil and gas discoveries in the Norwegian segment of the North Sea. Phillips Petroleum was one of the leaders of this exploration work. Oil and gas are now a substantial source of foreign exchange for Norway. Fortunately for Norway, these resources, relative to their current rate of production, are quite large, and are likely to quite easily carry Norway safely into the 21st century, in terms of having considerable petroleum to sell. By that time, perhaps the further development of hydroelectric facilities will provide Norway with another important energy export.

Peru. This country has a fairly broad spectrum of minerals, including petroleum, copper, lead, zinc, silver, and gold. Indeed, the gold and silver were the undoing of the Inca Empire, because Pizarro and his troops came to plunder the Inca treasury. Peru today is a poor country and much of the population is at a bare subsistence level. What is exported is largely minerals, which make up more than half of foreign exchange for that country; but no one mineral makes up a majority of the total. Of the minerals, copper is the most important. Peru's economy is in very difficult straits, with a sizable foreign debt (actually now essentially repudiated). Without mineral export income, Peru's economy would be in a disaster condition. It is close to being so now, and the poor population continues to increase. Minerals export is vital to that country.

South Africa. This country has been described as a mineral treasure storehouse. The description is valid, more about which is discussed in other sections of this book. Suffice it to say here that most of South Africa's export income comes from the sale of a broad spectrum of metals. No one mineral accounts for as much as half of that country's exports (gold amounts to 40 percent). Collectively, however, the metals and diamonds of the Republic of South Africa earn nearly three-quarters of that country's foreign exchange. The metals include gold, chrome, copper, manganese, tin, iron ore, silver, and platinum.

Zimbabwe. On the 18th of April 1980, Rhodesia (Southern Rhodesia) became the Republic of Zimbabwe. Like Zambia and Zaire, Zimbabwe is one of the geologically fortunate African countries, whose territory includes some of the rich mineral deposits of southern Africa. Indeed, Zimbabwe has, within its borders, what was called the Great Dike of Rhodesia, a unique geological feature which is a great wall or dike of rock rich in chromite. The dike is three to four miles wide, extends vertically to an unknown depth, and is more than 300 miles long. There is nothing comparable, as a mineral deposit, known anywhere else in the world.

Zimbabwe also has substantial deposits of gold, copper, nickel, and iron, which together make up a substantial part of the country's exports. However, it is also a rich agricultural country and exports tobacco, sugar, and other plant products. Nonetheless, for hard currency, mineral exports have been Rhodesia's and now are Zimbabwe's mainstay and are likely to continue to be.

Finally, in discussing one resource nations, in terms of their export earnings, it is interesting — and perhaps surprising — to note that the Soviet Union, despite the possession of perhaps the widest spectrum of mineral resources of any nation, falls into the category of a one resource nation in international trade. Substantially more than half of the USSR's foreign exchange, which it so urgently needs, comes from the sale of petroleum, chiefly natural gas, which makes up nearly 80 percent of the value from its exports.

The USSR has the world's largest, presently known natural gas reserves. But it does not have great fields of surplus grain, as does Canada, which supplements Canada's substantial export of minerals. Nor does the Soviet Union have the great timber resources of either the United States or Canada.

What grain and timber it does grow, it must use domestically. Only gas is currently in substantial surplus. Soviet oil production is now nearly in balance with its internal consumption, so the USSR, in terms of exports, is now rather markedly a one resource nation. The Soviets realize this and are trying to shift their power sources from the current heavy reliance on petroleum (oil and gas), more and more to coal and to hydroelectricity, reserving their petroleum, as much as possible, for sale, to obtain badly needed foreign exchange. However, in the future, the USSR may be better positioned in terms of a variety of mineral exports, for they are slowly developing the great mineral deposits of Siberia, which heretofore have been locked up by their remoteness and general lack of access. But the increasing network of roads and the new Siberian railroad (the Baykal-Amur Magistral) will perhaps allow the USSR to become a major mineral exporter in the 21st century. The energy mineral and mineral resource supply source scene constantly changes.

From this review, it can be clearly seen that many nations are markedly, and in some cases almost entirely, dependent upon one or a combination of several mineral resources for their livelihood, beyond a mere subsistence existence. These minerals, however, are a one-crop, non-renewable resource. When they become depleted to the point where they are no longer a significant source of national income, what will happen then? In many, if not most, instances, there is nothing else presently apparent to replace this loss of export earnings.

The history of numerous smaller communities and regions dependent on a given mineral resource was documented in the previous chapter, *Mineral Microcosms*. Will the rise and fall of such communities eventually be re-enacted on a larger scale by these single resource nations? In any event, it seems clear that minerals do now — and will continue to — influence the destinies of many nations. For some nations, minerals will be the dominant influence in their future.

Chapter 8

Mineral Riches and How They Are Spent

One of the interesting, important, and long-lasting effects from the discovery and development of mineral deposits has been the way in which individuals, communities, and nations have spent this new wealth. Once mined or produced from a well, a mineral or energy mineral deposit is gone forever. Oil fields and copper mines do not grow again. But the legacy of these resources, now gone, can be found in many forms, in many parts of the world. Some will be with us for decades (if not centuries) to come; others are more transitory. Thus, the legacy is mixed; and it is mixed further in that in some cases it will be of great and lasting benefit to people and to regions; but in other instances, it may literally mean that more people in the long run will suffer greater hardships, as populations expand on a resource base which eventually becomes depleted. It may not be unlike the over-grazing of a once rich grassland, which ultimately can support only a fraction of the earlier population.

It is the miner or the oil wildcatter who discovers the mineral. But they are not necessarily the ones who will make the largest profit. Frequently it is the entrepreneur who survives and reaps the greater rewards. Miners come and go; but the general store and the storekeeper selling the potatoes, flour, nails, and shovels stays on. Such was the case of Leland Stanford who supplied the California '49-ers with their basic needs. The miners are lost in history but Leland Stanford survives as the founder of a great university. The gold of the Sierra Nevada has largely been mined out; but its legacy, through Stanford University, in the form of educated human minds, is the most important resource of all. Thus a nation *can* benefit greatly from the discovery of mineral resources, long after those resources are gone.

It is reported that students going to the chapel at the University of Chicago, when they sang the doxology, would sing "praise John from whom oil blessings flow." It was John D. Rockefeller, with his oil money, who established the University of Chicago. At this same prestigious university, from a laboratory underneath the stadium, emerged the atomic bomb, which ended World War II.

Not nearly so well known was Rockefeller's partner in Standard Oil, Henry Flagler. He used his oil money to open up Florida, pushing a railroad through the State to Key West, building hotels along the way. He also bought huge tracts of land and lured people to Florida, both to live and as tourists. Oil money helped to put Florida on the map.

Andrew Carnegie built U. S. Steel Corporation, drawing upon its iron mines in Minnesota and Michigan. In small towns all across America, there remains a solidly-built, familiar structure, the Carnegie Library, some 2500 of them, ultimately originating, in effect, from the iron ore deposits of the Upper Great Lakes region. True, these libraries are small and for the sake of economy, most of them are of one distinctive design, but for many towns this was the start of bigger libraries, as communities outgrew the original structures. In many small towns, this was initially, and for many years, the only library for miles around. Who knows how many youngsters received their first introduction to science and literature in these places, and how many went on to have distinguished careers, which grew from this seed.

The Carnegie Endowment for International Peace is another legacy from iron ore. This fund, in a variety of ways, seeks to promote world peace, and has been active in the cause from the time it was set up to the present day.

The Carnegie Museum in Pittsburgh, the Carnegie Institutes in both Pittsburgh and New York, as well as Carnegie Hall in New York, are additional and prominent evidence of what American iron ore did for society at large. Although the high grade ore which made these things possible is now depleted, the legacy lives on.

This iron ore legacy even reached the Old World, for Andrew Carnegie was a loyal Scot, and finally retired to Scotland, where, among other things, he set up the Carnegie Trust for Universities of Scotland. Altogether, Carnegie gave away some $350 million dollars, at a time when the dollar bought much more than it does now; and he did it with the philosophy that "It is a disgrace to die rich." He was a generous Scot, who could be so by the generosity of geological processes, which took place more than a billion years ago, in the Upper Great Lakes Region of the United States.

A Rhodes Scholarship is one of the highest honors and most valuable awards a student can receive. By means of this fund, some of our most distinguished leaders have obtained their higher educations. These scholarships were set up by Cecil Rhodes, who conceived the idea for what was to become the British South African Company, and were established to develop the areas which ultimately became known as Northern and Southern Rhodesia (now Zambia and Zimbabwe). These were regions which were probably the greatest gold fields of the ancient world. But there was still enough gold left in Rhodes' time to produce tremendous wealth. The gold was found in an area some 500 miles long and 400 miles wide. The mines in ancient times had produced large quantities of gold; but sufficient gold remained in these earlier mines, so that about 90 percent of the approximately 130,000 registered claims were staked on the sites of ancient workings. Rhodes, through the South African Company, brought in settlers, chiefly miners, in the decade of the 1890s; and by 1905 this area was once again a world-class gold-producer.

Rhodes made a modest fortune from his interest in the South African Company; and in his will, he provided money for about 200 scholarships for a term of three years each, on a perpetual basis. The South African Company, based largely on the gold of that region, went out of business in 1923, as gold deposits were depleted; but the gold's legacy will survive, as long as Rhodes Scholarships exist.

The Guggenheim family derived their several individual fortunes from the silver of Leadville, Colorado, gold in the Yukon, copper in Alaska, Utah, and Montana, and diamonds in Africa, among other mineral resources. The several branches of this family set up a number of foundations, and made a variety of charitable contributions including $12 million to the Mayo Clinic and $22 million to the Mount Sinai Hospital; and established art museums and a free dental clinic. However, the most famous of the foundations is that set up by Simon Guggenheim in 1924, in memory of his son. This, the John Simon Guggenheim Memorial Foundation, sponsors Guggenheim Fellowships, which go to artists, scholars in the liberal arts, and scientists. Now more than 10,000 persons have benefited from this endowment, giving encouragement, particularly to promising young men and women, including such persons as Linus Pauling, who gives much credit to his Guggenheim Fellowship for helping him out at an important time in his life. He subsequently went on to win two Nobel prizes.

Rhodes, Carnegie, Rockefeller, Stanford, and the Guggenheims stand out as individuals who more or less single-handedly left large cultural monuments. But thousands of people, whose identities have been lost in history, have also built lasting monuments by their collective contributions. With mining or oil boom towns, come families, and the women, particularly, tended to express a desire for something besides saloons as cultural centers. And so it was that the opera first came to San Francisco, as a result of the gold rush. (The Alaska gold rush also brought opera briefly to that northern land, but the opera did not stay.)

With the major discovery of oil in Alberta after World War II, when LeDuc Number 1 was brought in, both Edmonton and Calgary greatly expanded, particularly the latter city, and the Calgary Philharmonic Orchestra, now one of the premier orchestras of Canada, is housed in a beautiful facility, paid for largely through the prosperity that oil brought. The back of any Calgary Philharmonic Orchestra program, where the benefactors are listed, is a quick way to find out which oil companies have offices in Calgary.

The Province of Alberta has wisely managed its public oil revenues and set up the Alberta Fund, from which bequests are made each year to a variety of cultural and artistic organizations. It also provides various financial advantages to Alberta residents with respect to taxes and loans. Presumably, if carefully managed, this fund will survive long after oil.

Alberta also built a magnificent structure to house some of its more spectacular ancient residents. In the striking badland topography, on the northwestern edge of the old coal mining town of Drumheller, about 70 miles from both Calgary and Edmonton, stands the world-class Tyrell Museum of Palaeontology. There is none better anywhere, and the star attractions, the dinosaurs, are wonderfully well displayed. In asking about the funding for this facility, a reply from the museum's director, Dr. Emlyn H. Koster, put it this way: "As one of the more costly cultural facilities of the Alberta Government, the Tyrell Museum capital project coincided with an affluent early part of this decade due to a very buoyant oil and gas industry." It was money very well spent.

In many small grants, and some not so small, minerals have contributed to institutions of higher learning. It is a rare university and college in the United States and Canada which has not received some money, some equipment, or some scholarship from a mining or oil company. With an endowment fund of nearly $3 billion, chiefly from oil revenues, the University of Texas is the world's richest educational institution. To fill the very fine campus buildings, the University has purchased (The term is used advisedly.) faculty from great universities world-wide. And why not? But now with oil revenues declining in Texas, it will be interesting to see if that institution can be maintained as it has been in the more opulent oil-filled past. One hopes it can be, inasmuch as the foundation for the continued good future of this fine university has been laid in the form of a large endowment fund which grew as the oil revenues came in. The trustees of the fund are well aware that oil is a one-crop situation and have been working hard to invest the money wisely.

In the publishing field, still another legacy came from minerals. A substantial part of the Hearst family millions was derived from the silver-rich Comstock Lode, discovered in 1859 at the present site of Virginia City, Nevada. Here George Hearst of San Francisco initially made a fortune. He did the same thing again (1877) from the gold strike in South Dakota. With this money, he established and expanded the Hearst publishing empire.

Mineral wealth not only funds institutions, and endows states and provinces, but in a number of instances, it has transformed entire countries. Brunei not many years ago was a hot, moist, small country, on the northwestern coast of Borneo. The population lived largely on a subsistence level. Then oil was struck. Now Brunei is a rich, hot, moist, small country, the citizens of which enjoy a per capita income among the world's highest — more than $20,000 a year. However, this is somewhat misleading, as a considerable amount of the oil income apparently goes to the Sultan. Nevertheless, because of employment in the oil fields and increased secondary business from the presence of the oil industry, this oil wealth has reached the general public, which now enjoys more than one car per family. The government does provide almost totally free education and medical expenses. There is also a generous welfare system, cheap subsidized food, and even subsidized television sets. For those of the 85,000 workforce who work for the government (and 46 percent of them do), there is a free pension system and very low interest loans, by which to buy cars and houses. No citizen pays income taxes.

How much the Sultan gets apparently is a state secret. *Fortune* magazine said that the Sultan was the world's richest person, with assets of $25 billion. Another independent estimate puts the figure at $28 billion. Those who manage the Sultan's money dispute these statements but admit that Sultan Sir Hassanal Bolkiah's personal fortune does put him in the billionaire class. (The King of Saudi Arabia is a distant second in the list of rich). The Sultan and his large family reside in a palace with 1,788 rooms, costing $500 million. The throne room seats 2,000 people and is hung with twelve two-ton chandeliers.

The monetary reserves of Brunei are estimated to be between $25 and $30 billion, or more than $100,000 for each Brunei citizen. This money is held for the good of the country and presumably is being, and will continue to be, put into sound investments.

The Brunei Investment Agency officials say they avoid common stock, but buy money market instruments, currencies, and property. The Sultan himself bought the Dorchester Hotel in London and the Beverly Hills Hotel in California.

Brunei has changed; but in terms of a tremendous transformation brought about by oil, Saudi Arabia surely stands out as the most striking example. Before oil was discovered, the country's economy was largely simple oasis agriculture, together with some fishing and pearl diving in the Persian (Arabian) Gulf, and nomadic herding of sheep and camels. Some additional income was provided by the pilgrims who came to Mecca. But the discovery of oil dramatically changed that nation, as no other nation has been changed in such a short time. This transformation has been described briefly in Chapter 9, *The Luck of the USA and Saudi Arabia*. More of that remarkable story is included here.

The Saud family united the various tribes of the Arabian Peninsula into the Kingdom of Saudi Arabia in 1932. The Port city of Jiddah, as recently as 1940, was a walled city of about 50,000 inhabitants. Jiddah is now a thoroughly modern city, including plazas decorated with sculptures, high-rise apartments, wide boulevards, and the tallest building in the Middle East, the 44-story National Commerce Bank. At one time during the development of all this, there were 355 ships waiting to unload cargo at Saudi ports.

As late as 1962, there was only one radio station in all of Saudi Arabia. The first railroad did not reach the Capital, Riyadh, until 1951. As late as 1950, there were no paved roads anywhere in Saudi Arabia. By 1982, there were more than 7,000 miles of main highways, and more than 6,000 miles of paved secondary roads. More than 15,000 miles of earth-surfaced rural roads had been constructed to some 7,000 villages which for the most part previously had never seen a road.

The International Airport at Jiddah, constructed at a cost of four billion dollars, covers more than 40 square miles, and is half again as big as Kennedy, La Guardia, O'Hare, and Los Angeles airports combined. The Jiddah airport was, at the time of completion, in terms of area, the world's largest, and is capable of receiving as many as 100 aircraft an hour. There are four grades of terminals here, the most lavish of which is the one designed for the royal family, the Royal Pavilion. It has a copper roof, white marble walls, and its main reception hall has Thai silk wallpaper and gold embroidered tapestries. Adjacent is an ultra-modern press room, with complete radio and television equipment.

But the airport at Jiddah was soon overshadowed by the airport built 22 miles north of Riyadh. This facility is more than twice the size of the Jiddah airport, covering an area of 94 square miles.

As late as 1978, there were only about 125,000 telephones in all of Saudi Arabia, but by 1985, in just seven years, there were more than a million phones in operation, serviced by equipment supplied by major communications companies in the Netherlands, Sweden, and Canada. Saudi Arabia today has the most advanced computerized telephone exchange systems, overall, of any country in the world. The reason for this is that the entire installation has only just been built, and there is no older obsolete equipment in service. As part of the communication system, Saudi Arabia has a domestic satellite system of 11 mobile and three fixed Earth stations, which allow almost anyone with a telephone, anywhere in the Kingdom, to dial directly to anywhere in the world. For a villager who 30 years ago had no roads, no lights, no sewer, no telephone available anywhere in the village, and spent the day herding sheep or camels, this is quite a change.

Hospitals and schools, up through the university level, have also been constructed. The Saudis have spent vast sums of money to speedily bring themselves into the twentieth century. Although some inefficiencies are inherent in such a mammoth and rapid undertaking, for the most part the money has been employed wisely, for even the Saudis know that the world's largest oil reserves are not infinite.

Along with the production of oil and development of large shipping facilities for this product, Saudi Arabia has realized that, rather than being just a raw material supplier for foreign interests, it should try to process more and more of its petroleum into finished products at home. To this end, it is developing a large petrochemical complex, located in the newly developed cities of Jubail and Yenbo, at a cost of about $45 billion, and using the talents of more than 40,000 people, drawn from 39 countries. These facilities may ultimately rival those for which the Houston, Texas area has long been known. This is inevitable, and the shift in emphasis from being a raw material producer to upgrading the resource to the finished end product will have a large economic impact on other such facilities around the world. The Saudis have the oil and gas reserves necessary to support these plants for many decades, if not a century or more to come, or at least long after the reserves of the United States and of the North Sea (Britain and Norway) have been essentially exhausted.

The shift of a raw material source to another country also means the loss of the plants which upgrade the raw materials; and with these plants go jobs, as well as the peripheral smaller suppliers and related jobs supporting these major processing facilities. Losing the capability to produce raw energy materials or metals is much more important to a domestic economy than the simple percentage the raw material itself represents in the gross national product. This point seems to have been largely unrecognized by the politicians and the public at large in the United States, where there has not been a great deal of general concern for the oil and mining industries. Indeed, these enterprises, fundamental to the economic well-being of the country, have been treated rather unkindly at times. But in the meantime, Saudi Arabia and more and more other petroleum producers are putting up refineries and other processing plants in their lands. Some metal producers are doing the same thing (Chile for copper, and Venezuela using its gas to set up aluminum smelting facilities for its own ore and that of neighboring Suriname).

Water has been a limiting factor in the economic life of Saudi Arabia, but petroleum has come to the rescue there, too. Using the abundant natural gas supplies available along with the oil, the Saudis have developed desalinization plants now producing more than a billion gallons of fresh water a day from the sea. Using waste heat from these plants and other facilities, all based on petroleum, electric power is generated. The story of the electrification of Saudi Arabia is perhaps the most spectacular of all.

The first public electric generating plant was opened in Taif, in the late 1940s. Since then, nearly the entire country has been electrified, and all major buildings are now air-conditioned. So, in 50 years, Saudi Arabia went from nomad tents in the desert, to air-conditioned, multi-story office buildings. There probably is no greater transformation of a country of such size, anywhere in the world, than that seen in Saudi Arabia in the past five decades. Without petroleum, this region would have remained simply an obscure arid land.

One of the interesting touches which oil brought to Saudi Arabia is a 60-ton, solid granite bathtub. In 1983, King Fahd ended his world-wide hunt for a perfect bathtub in a granite quarry in Manitoba. A flawless piece of granite was cut and hauled to Montreal and then shipped to Italy, where some of the world's finest stone workers cut, sculptured, and polished it, including the final touch — the royal family crest. It was then shipped to Riyahd and installed in the palace.

Mineral bonanzas can cause some quite unexpected things to happen in a national economy. In the case of Saudi Arabia, its huge petroleum deposits ultimately produced a glut of wheat. It all began in the late 1970s, when the Saudis thought that the wheat producing countries might retaliate for oil shortages by cutting off wheat shipments. So the Saudis embarked on a program to encourage domestic wheat production. This was done by means of a subsidy, which promised farmers a price of $26 a bushel. The program was so successful that in a few years the local production of wheat had reached two million tons, whereas domestic consumption was only one million tons. The Saudis cut the subsidy price to $14 a bushel, still five times the world price, so the wheat continued to pour in. When the government tried to reneg on its commitment to buy the wheat, the farmers raised such a storm that the government had to pay up. The problem continues, and what is perhaps more serious is that Saudi agriculture, including wheat in particular, requires 84 percent of the water used in the country, and 70 percent of that comes from underground aquifers, which are not now being replenished adequately. The water table is dropping. They have been mining their groundwater to produce a crop they do not need and paying five times the world price for it. Striking oil can cause some odd complications.

Another facet of the Saudi oil bonanza has been the training of Saudi young people. At one time, virtually any male (women were generally excluded), who could qualify academically, could study abroad at government expense, and study almost any subject he wished. The effect of this has been to produce a flood of well educated people. But as the oil boom moderated, particularly during the 1980s, when oil plunged from nearly $40 a barrel, briefly, to less than $10, oil production was cut and economic activity diminished. The result has been to cause a substantial number of recently educated young Saudis to become unemployed or underemployed. In this group may lie the seeds of political problems in the future. Some of the new-generation Saudis are reluctant to accept the kinds of work they are asked to do. Foreign workers have been doing many of the more mundane tasks but as oil income decreases, given more moderate prices now, these foreign workers who drained money out of Saudi Arabia, much to the benefit of their home countries, are being requested to leave. The young Saudis are asked to take their places. Many of these young people come from the proud Bedouin tribes, and do not have a manual or technical trade work ethic. They expect to have offices, secretaries, titles, and authority. They do not want to work with their hands, or on a farm. So far, the government has been able to keep these people reasonably content with grants and loans for houses, free schooling for their children, and monthly stipends for continued support of students abroad.

By 1985, however, the Saudi budget was in the red, as expenses exceeded oil income; and they began to live off of their accumulated capital. The affluent society, so rapidly and spectacularly built up by oil income, faces more austere times; and the long-term effects of this will profoundly affect the Saudi Arabian political structure. Can it survive as a kingdom, in the face of a growing and possibly discontented middle class? To keep things going, the Saudi treasury has had to draw down its reserves, which went from $141 billion, in 1982, to $109 billion, in 1986. And the kingdom's budget manager began to use a month to month funding procedure, because projection of oil revenues a full year ahead could not be made with any degree of certainty.

Another result of the decrease in oil revenues, which fell from a high of $100 billion in 1980 to only slightly more than $20 billion in 1986, was what one observer called "the rediscovery of religion by the sheiks." Various Saudi enterprises had expanded greatly during the oil boom years and borrowed much money to do this. Islamic law precludes the payment of interest on loans; but money borrowed produced such a good return that the Arab businessmen did not mind the banks charging interest, in effect, by calling it "administrative fees," or "loan initiation discounts." And as prices of everything roared upward and business prospered, nobody cared; but when things turned around and headed down, the loans became a problem. The borrowers then began to invoke "sharia," the Islamic law which prohibits the payment of interest. Furthermore, the general nature of Islamic law, and the judges who administer it, tend to markedly favor the debtor over the creditor, and in Saudi Arabia, sharia is the law. This problem started in Saudi Arabia but subsequently spread to adjacent Muslim oil-rich countries, such as the United Arab Emirates, and Bahrain.

Foreign banks have found they can do little about this matter, for not only can they not collect interest; but Islamic attitudes essentially preclude seizing the debtor's assets, even though they were originally pledged as collateral for the loans. The Arabian Auto Agency owed $300 million to 43 international banks. The Arabian Homes Company owed banks in the western world nearly $80 million. The literal translation of "sharia" is "path to the watering place," and, as one report duly noted, the foreign banks are taking quite a "bath."

Interestingly enough, however, it turns out that whereas being involved with interest on money in Saudi Arabia is against Islamic law, it has not hindered quite a few Saudis in investing their money abroad in places paying interest. This shipping out of money to foreign interest-paying investments, with the thought that perhaps Allah will not find out, or at least overlook it, is not officially approved; but it is widely done. As a way of accommodating Allah's economic rules legally, the Saudi banks have set up overseas mutual funds that pay their interest in the form of additional shares, which then can be sold, the profits not being regarded as interest (Allah never thought about mutual funds.). And domestically, the Saudi banks are thinking about issuing zero coupon bonds, to be sold at a discount from face value. Again, Allah never considered this route around the no-interest rule either. The love of money does great things, even in religion.

On a smaller but equally lavish scale, the story of Saudi Arabia and what oil did for it has been repeated in Kuwait. This was a British Protectorate until 1961, when Kuwait was given full independence. It has an area of only about 7,000 square miles, and lies at the northwestern end of the Persian (Arabian) Gulf. Prior to the discovery of oil, it was an insignificant piece of sand whose principal exports, such as existed, were skins, wool, and some pearls. The discovery of oil changed all that. Beneath that desert area lies an estimated 91.9 billion barrels of oil yet to be produced. With 7,000 square miles, Kuwait is smaller than some counties in the United States (for example, Harney County, Oregon, 10,166 square miles); but the proved reserves, that is, oil already discovered, are nearly four times the reserves of the entire United States including Alaska, with a total area of 3 1/2 million square miles.

With the discovery and development of oil came free education for all Kuwaiti citizens as far as they want to go, no income taxes, subsidized housing and subsidized utilities, free medical care; and the government pays more than $7,000 to every couple upon their marriage. The country has its own airline, Kuwait Airways, which flies Boeing 747s on regularly scheduled service from New York and London to Kuwait. Among the numerous minor facilities built by the Kuwait government there is included an Olympic size ice-skating rink.

To develop Kuwait as rapidly as has been done, a great number of workers were imported. At the peak of activity, 75 percent of the work force was foreign. The foreigners out-numbered the native population. There is obviously something of an internal threat in this situation; but it has merit also, in that when times are slow economically, the country can simply export its unemployment problem by decreeing that the foreign workers return to their homelands. More recently that is what is being done.

There are only about 600,000 Kuwait citizens among which to divide the income from the estimated third largest oil reserves in the world. Understandably, it is very difficult now to become a Kuwait citizen if you were not born one. Rich as it is at present, Kuwait has a keen eye on the future, for oil, as they say, is a "soft loan from Allah; use it wisely." Since gaining independence in 1961, Kuwait has invested its large oil revenues into a great variety of assets. In the top 70 blue-chip stocks listed on the New York Stock Exchange, Kuwait has invested at least $2 billion. Kuwait also has set up what is called the Kuwait Fund for Future Generations. This organization and the Kuwait Petroleum Corporation have bought 14 percent of the West German car-maker, Daimler-Benz; 20 percent of the German mining complex, Metallgesellschaft; and about a quarter of the giant German chemical company, Hoechst.

Perhaps endorsing the expression, "there will always be an England," even after oil, Kuwait bought 5.1 percent of the British Midland Bank and has also invested heavily in English real estate. They own a one million square-foot complex of restaurants, shops, and offices along the Thames in London; and they own about nine percent of British Petroleum Corporation. Kuwait has put more than $2 billion into Madrid, Spain and is looking at possible investments in Australia, Singapore, Malaysia, and Hong Kong.

In the United States, Kuwait has purchased Santa Fe International, a California-based drilling company; and they also have a very large portfolio of United States real estate, including 100 percent ownership of the Kiawah Island Resort in South Carolina. They own several New York's skyscrapers but have kept this fact obscure, in deference to the Jewish tenants in some of these buildings.

In order to assure a good market for its oil, Kuwait has bought most of the European marketing and refining operations of Gulf Oil; and they are adding to this system by building more gasoline stations of their own. The Kuwaitis have a brand of gasoline they market as Q8 which is distributed through more than 4,700 retail stations in a number of countries which include Italy, Denmark, Sweden, Belgium, the Netherlands, and Luxembourg.

Recently, Kuwait has been earning more money from these other enterprises than from the export of its oil. When the oil is gone, Kuwait hopes and believes that these investments will continue to allow them to pursue the affluent life-style they now enjoy. Time will tell; but clearly Kuwait is very conscious of the finite nature of its oil riches, and has been working very hard, and quite successfully, to put itself in a position to survive beyond petroleum. A Kuwait government economist states "Our investments will be the main source of income for generations after the oil runs out."

However, in late 1989, Kuwait announced it was drawing up plans for a multibillion dollar petrochemical industry, to upgrade its end product, rather than just shipping out crude oil.

Dubai, of the United Arab Emirates, in six-page ads in American business magazines, quotes Sultan Ahmed (in Sulayem) saying "We knew the oil boom would not last forever, and that a more stable and diversified source of capital was required." Dubai's answer to this eventuality has been to create what they like to call "The Hong Kong of the Middle East." This is a mammoth, free-trade zone, which first required that they dig a 2,500 acre harbor at the edge of the desert. Around it, on 25,000 acres, an industrial and warehouse complex has been built.

The concept is that the United Arab Emirates' location, at the lower end of the Persian Gulf, gives them a geographic advantage in being the trans-shipper of goods from the Middle East into Africa and Asia. The free-trade zone imposes no inconvenient tariffs, and offers a number of other financial advantages. Adding to this, Dubai now has a first-class airport and its own international airline; and four hospitals, a golf course, a $14 million cricket and hockey stadium, tennis and squash courts, and a bowling green have been built. They also advertise that beyond golf, tennis, and the like, you can watch camel racing from a special viewing building. They further note that "One popular pastime pursued in Dubai, that is found nowhere else in the world is, 'wadi-bashing.'" Its rules are deceptively simple. It involves nothing more elaborate than driving in wadis, which are dried up river beds that flow from the Hajar Mountains into the desert, or, on the East Coast, toward the sea. So far, this "free-trade zone" project seems to have been a success.

Algeria, torn in 1988 by riots over price increases, food shortages, and unemployment, nevertheless bought two Soviet Kilo-class diesel-electric submarines that year. Which brings up the matter of using limited, oil-derived resources (In the case of Algeria, it is mostly gas.), for military hardware, instead of the general benefit of all citizens. And we consider, Libya.

Libya is another example of a largely desert country, catapulted into the twentieth century by oil, and oil alone. Libya has an area of about 680,000 square miles, or slightly less than one-fourth the size of the 48-adjacent United States. At the time oil was discovered in 1950, Libya had a population of only about two million people. With merely a narrow strip of greenery along the coast suitable for conventional agriculture, the land, as an agricultural economy, could not support many more people. The impact of Libya on the world scene prior to the discovery of oil was slight indeed.

But now it has oil revenues from the production of about two million barrels a day.

Libya has had a mixed spending program for this wealth. Early after the discovery of oil, under the leadership of the wise old desert chieftain, Idris, who became king, money was spent on roads, sewers, water wells, and other generally useful public enterprises, to the benefit of a great many people. But Idris had to go to Greece for some medical attention; and while he was there, a Colonel led a coup to depose Idris. And thus Col. Khadafy came to power, and the spending emphasis shifted from civilian projects to military hardware. The ground equipment of 20 tank battalions, modern missiles, trucks, and artillery came mostly from the Soviet Union. In the air, French Mirage fighters appeared. Six ex-Soviet submarines and numerous small surface craft make up the Libyan navy. Some public works were continued, but the planning became a little haphazard. In part of a major road system, where clover-leaf interchanges were constructed, apparently the project was less than well thought out, for at several places it was necessary to make a U-turn against on-coming traffic in order to get on another road.

But such things are of lesser concern; for with the aid of oil revenues, Libya has been engaged in exporting terrorism and invading neighboring Chad, to the south. In 1989, Libya was found to have built a chemical weapons plant. Without oil, Libya would still be on the fringes of the twentieth century, looking in from the outside; and it would continue to be mainly a country of nomads and small farms. What Libya has invested in, to keep it going beyond the time of oil is not very visible. Libya has made some modest, long-term investments, including, for example, a one-half interest in an oil refinery in Hamburg, Germany. They also own a small airline (Libyan Arab Airlines — LAA).

Recently, a grandiose scheme has been announced to pipe water from deep below the sands of the Sahara to the coastal areas. Khadafy says that his Great Manmade River Project will convert the country (which is now 95 percent desert) into a "garden of Eden." The plan calls for some 12,500 miles of pipeline to carry water to all areas of the coast. However, as the wells are being drilled in a desert area where there is no apparent very rapid recharge, it is likely that the water has accumulated over a long period of time. Any heavy pumping of the aquifer will exceed the recharge and water levels will drop markedly. The long term success of this project is doubtful.

For the most part, from what is visible at the present time, it seems likely that tanks, guns, planes, and second-hand Soviet submarines will be a large part of the legacy which future Libyans will inherit from their once-affluent oil era. But currently continuing to pursue this course of spending, in early 1989, Libya purchased several Su-24 Soviet bombers. Again one has to wonder about the logic behind a desert country, of little more than four million people, basically poor except for the temporary oil riches, spending its money on military planes. The judgement of history on the present way in which these transitory oil riches are being spent is likely to be severe.

Before leaving the oil-rich countries of the Middle East, including Libya, in this discussion, there is one major expenditure which appears to over-shadow all others; and our recent citation of Libya's military purchases illustrates it. In the 12 years following the first oil embargo, which occurred in 1973, with the subsequent huge rise in oil prices and concurrent great increase in revenues to these oil producers, Gulf nations spent $640 billion for military purposes. Iraq and Iran spent $220 billion between them, and were able to finance an eight-year war of attrition (which ended in a stalemate), with an estimated one millon people killed and probably twice that many wounded. Without the oil money to finance the war, it is probable that at least the casualty figures would be somewhat less. Oil has been a mixed blessing to these countries.

Now the Gulf nations have the latest state-of-the-art weaponry, including all sorts of missiles, tanks, jet fighters, helicopters, warships, and other hardware. But there is a question as to whether or not these nations are happier with all these devices of destruction.

Mexico is another country which had the good fortune of having rich mineral resources. However, initially this may have actually been a curse, for it brought in the Spaniards, who plundered and ultimately destroyed the Indian civilizations there, as well as smallpox and other diseases, which decimated many of the native peoples. The mineral resources which Cortez and his cohorts sought were gold and silver; and Mexico remains the largest silver producing country in the world. But more recently, the black gold called oil has become the most important mineral resource in Mexico. Oil was discovered quite early in Mexico but only in modest quantities, although what was probably the world's largest single oil well was drilled in Mexico, producing some two to three million barrels of oil a day, for a time.

But it wasn't until the second half of the twentieth century that the great oil fields of the Tampico-Vera Cruz area, especially offshore, were found. The result is that Mexico now has larger proven oil reserves than does the United States. Unlike in Saudi Arabia, however, the oil of Mexico has not done a great deal to transform the country. True, it has built the tallest building in Mexico City, the Pemex Building, and it has brought local prosperity to some of the coastal communities, where the oil fields are located; but for the vast majority of Mexicans, the oil has had little effect, except perhaps to keep the price of gasoline from being as high as it might be, if Mexico did not have the oil resources. But how much of the oil money has gone into increasing the Mexican standard of living is questionable. Ask a taxi driver in Mexico City and he will sigh and say that the money has gone mostly into the pockets of politicians; and substantial sums have fled the country for safer havens abroad. Good or bad, this was indeed a strategic move by whomever had the money, for the Mexican peso in the 1980s went from about 12 to the dollar to over 2,700 to the dollar.

An anomalous result of the big Mexican oil discoveries may be that Mexico will actually be deeper in debt because of them. On the basis of the oil finds, foreign banks, most of them American, were persuaded to loan large amounts of money, now totaling more than 100 billion dollars, to Mexico. Where all this money has gone is somewhat of an enigma, but what clearly remains is a huge foreign debt.

Nauru, the little Pacific island nation, has a single but very rich mineral deposit — phosphate (discussed in detail in Chapter 7, *The Chiefly One Resource Countries*). It also appears to be the nation that has the mineral resource with the shortest economic life-span. This phosphate deposit will probably be depleted by 1995, if not earlier. The challenge is to immediately put in place investments somewhere which will continue to support the population. To handle these investments, the country formed the Nauru Phosphate Royalty Trust. One such investment is Nauru House, a 52-story office building (the tallest) in Melbourne, Australia. The building is big enough to accommodate the entire population of Nauru (legal citizens number about 4,000), if they chose to convert it into a hotel.

Nauru also has its own airline, consisting of five Boeing planes. Air Nauru touches all major air terminals in the Far East, and is the only line to run a north-south route through the mid-Pacific region. Nauru has a shipping line, the Nauru Pacific, and two excellent fishing vessels. Looking toward both immediate and future tourist trade, Nauru built a hotel on their island. In early 1988, Nauru bought a square-mile tract of land in the West Hills district of Portland, Oregon, with the intention of making it a luxury home development.

Whether all of these and perhaps other investments in the very near future (before 1995) can become the permanent source of income for Nauru remains to be seen. It will be the first nation in the world to see its mineral resources on which it is mostly or wholly dependent, become totally exhausted, and move into the future on the basis of its investments alone. It will be an interesting situation to watch.

Nauru does issue very colorful postage stamps, which collectors seem to like. How long this source of income will last one can only guess.

Nigeria is the largest African nation in terms of population, and also quite recently has become a major oil producer and a member of OPEC. When the Arab nations cut off the oil supplies to the United States in the 1970s, it was fortunate that Nigeria at that time, rather than Saudi Arabia, was the principal foreign supplier of USA oil from the eastern hemisphere. However, Nigeria had no surplus then to make up for the oil which the Arabs did not supply and thus there was a shortage. However, without Nigeria's oil, the situation would have been much worse in the United States.

With its considerable oil income, Nigeria has embarked on a variety of projects, ranging from fine new government office buildings to steel mills. A grand scheme of national industrialization brought hope to many Nigerians, who flocked to these newly established industrial centers in search of jobs.

When President Shagari, who was from the northern part of the country, came to power in 1979, he had barely enough votes from the other regions to capture the office, so he had to do something to establish a political consensus for his regime. He did this by trying to spread the oil bonanza money throughout the country, in the form of large public-works projects. A side effect of these projects was that the percentage of Nigerians living on farms dropped to 65 percent from 85 percent, as people moved to the cities. And they also changed their tastes from home-grown millet, yams, casaba, and sorghum, to imported foods, so that some of the advantages of the oil income, in terms of providing foreign exchange, were lost, simply to fill the new demand for foreign foods.

International salesmen, seeing a pile of oil money, entered the scene and sold projects to Nigeria which, in some cases, were not very well suited to the economy, with the result that some have become economic disasters. Furthermore, once these public works and industrial projects are established, people tend to regard them as permanent; and when the oil income shrinks, keeping all of these projects alive becomes an impossible task.

Yet, given the political realities, it is likely that if the government of the moment is to survive, in some way these projects must continue. Thus, because of oil, an infrastructure is built which has no way of coming down without political and social problems, some of which can be severe. This is not only a problem in Nigeria, but a problem which many newly rich countries face in setting up projects and social benefit programs that cannot be sustained in the long run. Rising oil income can bring about political stability, social harmony, and a higher standard of living; as long as there is more and more to divide, people are happy. But just the reverse may also happen, when oil and other mineral income drops. Nigeria is just beginning to experience this latter situation. It will be watched with interest by other nations, which sooner or later must face the same problem.

However, Nigeria has been making efforts to invest some of its petroleum revenues in enterprises for the long-range national good. A recent example has been the completion of a world-class fertilizer complex at Onne, Rivers State, near Port Harcourt. The $800 million project serves domestic as well as export markets. With the over-cropped soil in many parts of Nigeria, this fertilizer will be most useful. The problem with this project, for the very long term, is that the feedstock for this plant is natural gas (from which the ammonia fertilizer is made), so it is still dependent on petroleum and cannot last beyond the life of petroleum. However, this project, which obtains the gas from the Aalakiri Field about nine miles away, will save Nigeria about $100 million a year in fertilizer imports, which will be a substantial help in the near future, and for a number of years to come. But again, as agriculture and population expand on this non-renewable resource, what is the ultimate outcome? Can the greater population continue to be supported in some fashion after the petroleum is gone?

Norway is a land of mountains, beautiful fiords, and lots of rock. Norway has always had to live, in considerable part, from the sea. It continues to do so from the oil and gas discovered in the Norwegian sector of the North Sea. The result has been that almost half of the Norwegian government revenues are now derived from petroleum.

How have these monies been spent? Norway has used much of it to keep unemployment low. Billions of dollars of petroleum revenues have been poured into what turned out to be money-losing projects in agriculture, iron mining, smelters, and fishing to keep people employed. Although this solves the employment problem temporarily, it does not provide a long-term, sound, economic base.

The question remains: Can these oil revenues be put to use in building permanently productive industries?

Norwegians have come to look at neighboring Sweden, which has only modest mineral resources and nothing to compare (in terms of income), that with which North Sea petroleum brings to Norway. Yet Sweden has developed some major international companies, that bring in a continual stream of income. That Norway has not done the same thing has been a frustration to some Norwegian economists, one of whom is quoted as saying "We're in such a mess. When are we going to realize we should hire Swedes to run our economy." For a Norwegian to say that surely indicates some basic problems in how to handle the new petroleum riches.

But Norway has begun to change its priorities in the way it is spending its oil money, and is putting more into research. Hallvard Bakke, Minister of Cultural and Scientific Affairs, has stated that "Our main objective is to expand the possibilities of the Norwegian economy and to give it more feet to stand on so that it does not have to rely on oil and oil-related industries." An example of this is the decision to pursue genetic engineering and apply it to one of Norway's new and rapidly growing enterprises — aquaculture. Money is being used to study the Norwegian fiords as a place to raise a variety of fish such as halibut, salmon, and cod; and perhaps by genetic engineering, produce some faster growing varieties which can be adapted especially to fiord life.

In review, we have seen mineral dependent countries such as Nauru, which almost immediately faces the fact of its one and only mineral running out; and Saudi Arabia, which has a mineral resource in petroleum that will last for many decades. But all of these countries have built their economies on a depleting resource. It is evident that some have spent their income quite wisely for the future. These are chiefly the countries which have a small population and a large resource. (for example, Kuwait and Saudi Arabia). They have had a surplus of cash from their oil income, which they could invest either within their own countries or abroad. Other countries, generally with larger populations, have had to use their mineral income to support current costs, and have not be able to put away any great amount of investment money for the future. Brazil has been in this situation to some extent. It has opened up large, rich, iron mines; but the iron ore it exports unfortunately has not been able to pay for the oil it has had to import.

Thus the legacy of how these mineral monies are used will run varied courses. Some run their course as soon as they are received — they are spent immediately, in one fashion or another.

But the mineral income money which is invested will reflect the longer term view and wisdom of the governments of today; and will markedly affect the lives of their citizens, probably for many generations to come.

On a smaller scale, the wisdom and foresight of those individuals who have had the good fortune of benefiting from mineral bonanzas, and wanted to share them, will also affect future generations through the universities, museums, scholarships, research foundations, libraries, and other beneficiaries of these gifts.

The high grade iron ore of Minnesota, now gone, and the oil of Kuwait, after it has all been produced, will leave their marks on many people in the future; quite a few of whom will never realize from whence their benefits came. Most of us today have had our lives touched, at least in a small way, by these mineral endowed situations.

"We have all drunk from wells which others have dug."

Chapter 9

The Luck of the USA and Saudi Arabia

Mineral resources have had a significant impact on the history and development of many nations; but two countries stand out in this regard — the United States and Saudi Arabia. Both are phenomena which almost surely will never be repeated, and are worthy of special note.

The United States went from a three million square mile area of raw wilderness to become the most powerful and affluent nation in the world — all in about 200 years. In terms of the total energy minerals and minerals spectrum, the United States was without equal at the time the Declaration of Independence was signed. However, the citizens of the time really did not know what riches existed in this new country; but as the pioneers moved westward and exploration proceeded, word spread of the great natural resources of this area, in both minerals and land. A flood of immigrants swept into this untouched territory; and in two centuries, the United States became the premier world power, with material wealth beyond anything ever seen in all of history.

The luck of the United States was that the country was established at the right time, at the right place, with the right people. In terms of the right time, the USA emerged as a nation shortly after the Industrial Revolution had appeared. Great Britain had started it, promptly to be picked up by Europe. New inventions and new technologies came on rapidly. But Great Britain and Europe had a lesser quantity and quality of mineral resources than did the new nation across the sea, and so when the immigrants of the time (which were largely from these northern European lands) came to North America, they were to find a much greater and richer spectrum of mineral resources than what they had left. They had come to the right place.

And they were the right people, because for the most part they came from lands where they had been under kings and oppressive landlords. Most of the people who came were not the nobility. Why should anyone who had it made in Europe come to the wilds and the primitive living of North America? A few did, but the great majority of the people who came were the down-trodden and those with little material wealth. But the majority were united by one thing — they had a burning desire to establish a country wherein all were given equal rights, regardless of their status at birth, and where what one gained by hard work one could keep. Thus was written the Constitution and the Bill of Rights, and what emerged was a democracy and the free enterprise, American capitalistic system.

This combination of the right time — coming of the Industrial Revolution — with the right place — three million square miles of virgin land, with a tremendous variety and quantity of mineral resources — together with a poor but ambitious free people, produced that miracle which became the United States, in just the first two hundred years of its existence.

At least one other country had the resource potential to replicate what happened in the United States. That country is the Soviet Union, which also has, as is now known, a vast and varied energy mineral and mineral base. But it lacked, two hundred years ago, the great influx of people who had the freedom to make for themselves in their new land what they had dreamed about — a political system which preserved their liberties, and an economic system which allowed them to accumulate wealth and invest it, in order to produce more wealth in the form of useful goods, for the broader population. The Soviet Union at that time already had a political, social, and economic system in existence; but it was not one to promote the sort of growth and vitality which has characterized America; and it still does not have such to this day. Probably it is easier to install a new political and economic system in an *unoccupied* land (e.g., the United States) than to change an already established system, such as existed in Russia.

One might suggest that Canada also had the same potential as did the United States. But Canada has a somewhat less varied mineral spectrum; and its northern position and more hostile terrains, as compared with the United States, delayed its development. Indeed, it was not until after World War II that substantial oil was discovered, with the advent of the LeDuc field and those which followed. Also, because of the more hostile terrains and a less favorable climate, in large parts of the land, Canada did not attract the population necessary to form the basis for a large industrial complex, with the sizable internal markets needed to foster large scale industrial development, such as occurred in the United States.

The United States has had an incredibly fast ride to the top of the world economic heap. It became very quickly the world's largest iron ore producer. And almost immediately thereafter, oil was discovered in 1859; and the United States became the world's largest oil producer, and was completely self-sufficient in petroleum for more than 100 years following. Ultimately, it was the possession of these large oil resources and the self-sufficiency thereon, which brought about the reversal of strength between Great Britain and the United States. Until World War I, coal was the dominant energy source. After World War I, oil became the major fuel by which the world progressed. Great Britain had the coal but the United States at that time had the oil, and with the arrival of the age of oil, the seat of power went *with* the oil, and *to* the United States. One might add, parenthetically that the current increasing dependence of the United States on foreign oil has substantially decreased the relative economic strength of the United States — a matter taken up later in this volume.

But for many years, the United States was the world's dominant producer of most vital raw materials. The United States was, until 1982, the world's largest copper producer, and still is the world's largest cement producer. Until about 1950, it was the world's largest oil producer. It has long been the leader in molybdenum and lead output. The possession and huge production of all these and other minerals helped the United States rise to its premier world position.

But the United States also benefitted from what can be only described as a uniquely favorable sequence of events: the timely combination of which created great forward momentum for this young nation. The geographic arrangement of some of the resources was also very fortunate. In this case, it was iron and the ingredients which go to produce steel. The richest iron ore deposits then known in the world were discovered in the Mesabi Range of northeastern Minnesota. The large, lower-grade taconite deposits had been fractured, weathered, and leached of worthless rock material, so as to leave behind the mineral hematite, which is 60 percent iron.

These rich iron ores were easily and economically brought together with the two other ingredients of steel-making, coal and limestone, by the fortunate geography of the northern Great Lakes region. This allowed the iron ore to be brought down first by rail (and it was downhill, not uphill, an economically important fact for the transport of heavy iron ore) to Lake Superior where cheap water transport brought the ore first to Chicago, where the first American steel rails were rolled in 1865, and then to the Pittsburgh area; both areas had the coal and limestone needed to combine with the iron ore to produce iron and steel.

This rich iron ore discovery came just when it was needed — at a time the railroads began to dominate transportation. The engines, the cars, and the rails all demanded great quantities of steel. The blast furnaces around Chicago and Cleveland and Pittsburgh produced it. American steel production was only 20,000 tons in 1867. By 1895, however, it had passed the British production of six million tons, and was 10 million tons before 1900. Ultimately, steel rails were to stretch from coast to coast, an impossible task then, had it not been for the great iron ore deposits which had been so timely discovered and developed.

Steel also built factories and machines by which more goods could be produced. The new railroads efficiently distributed the manufactured products, such as steel farm implements for the pioneers breaking sod in the Midwest and the Great Plains; the needed equipment to miners and ranchers of the mountain regions; and finally to the west coast.

Steel made the first skyscraper possible. After the great Chicago fire of 1871, large areas of that city were bare and in need of being rebuilt. An architect named William Jenney came up with the idea that walls of buildings would no longer be used for bearing the weight of the structure; but rather with the abundant and relatively cheap steel now available, he could build a steel frame to be the skeleton of the building. Then using lighter weight materials the structure could be walled in. Thus was born the first skyscraper, which was the 10-story Home Insurance Building, finished in 1885. It was such a success that two more stories were added later. Thus, the giant steel mills which came into being because of the rich iron ore deposits of the Mesabi Range, built our railroads and started the huge inner city complexes of office buildings as we know them today.

And the good fortune of the United States continued. About the time the steel business was booming, the electrical age was also dawning. Thomas Edison had produced the first usable electric light and visualized lighting cities with his invention. The electric motor had been invented about 1854. In 1879 Thomas Edison produced the first usable electric light and visualized lighting cities. But how could the electric current be transmitted to these lamps for use in homes, offices, and factories and to the motors which could replace so much hand labor drudgery in the factory?

Again, good fortune smiled on the USA with the discovery and development of huge, native, copper deposits, some of the richest known in the world, on the Keweenaw Peninsula of Upper Michigan. These deposits came into full production just at the time they were needed by the blossoming electric age. Copper became the workhorse metal for the electrical industry; and Upper Michigan, located not far from the industrial East and Midwest, where most of the copper was being used, produced huge amounts of this most useful metal. It was inexpensive copper — native copper. A tunnel in one mine struck a mass of pure solid copper about 50 feet long, with an average thickness of about 14 feet, and weighing more than 500 tons. The copper, being so malleable, could not be blasted out, but instead had to be cut into small pieces, a process which took more than a year.

The copper from Upper Michigan, made into thousand of miles of wire, carried electric power to light homes and factories. It made the workday more pleasant and efficient, and domestic life brighter. It converted the evenings into much more cheerful times than previously was the case; and it allowed factories to operate three efficient shifts a day, instead of one, if such production was needed — all adding to the much greater efficiency and productivity of the American economy.

In the 1830s, Samuel Morse set up his telegraph line from Washington to Baltimore. Soon copper telegraph wires spanned large areas of the nation, first running along railroad tracks, and later spreading out and connecting many otherwise isolated communities with the outside world. Later, telephones began to appear and again copper wires were available to put this most useful instrument into many places. Business and industry were greatly helped by this communication system. All this was facilitated by the abundantly rich copper deposits in nearby Michigan and developed at just the right time to promote the electrical age in the United States, in all its many and varied forms. Finally, it should be noted that the Michigan copper deposits fed far more money into the American economy than did gold from the California gold rush.

And then came the motor vehicle era — made possible by the abundance of cheap steel, combined with the discovery of oil in increasing amounts and in many parts of the United States. Starting first in Pennsylvania, drillers soon found oil in New York, West Virginia, Ohio, Texas, Louisiana, Oklahoma, Kentucky, Kansas, Colorado, Wyoming and California. Oil was found coast to coast; and by 1909, the United States was producing more oil than the rest of the world combined. In 1930, the great East Texas field was discovered, the biggest ever found in the 48-adjacent states; and oil prices dropped briefly to as low as 4 cents a barrel. So the United States found itself with more and more and cheaper and cheaper oil, and ultimately led the world from the coal era into the age of oil. Oil, in myriad ways, powered the United States to its dominant world position, including that of the world's largest automobile manufacturer.

Mineral and energy mineral resources were in seemingly endless supply and the United States successfully provided its allies with vital energy and mineral resources to first win World War I and then did it again in World War II. It has been said that both wars were fought out of the great hole in the ground which is the Hull-Rust iron mine, on the north side of Hibbing, Minnesota.

Blessed with cheap and abundant mineral and energy mineral resources, the United States enjoyed a phenomenally rapid ride to the world's highest standard of living and to being the world's greatest power — in just two centuries. The coincidence of the Industrial Revolution, combined with an energetic people and free economic and political systems, had much to do with this rise. But it could not have been so easily accomplished, and probably not accomplished at all, without rich mineral resources as a base on which to build the military and economic power it became. Iron ore, coal, and limestone for the steel mills, copper in great abundance for the dawning electric age, and oil to power the machines of industry and the wheels of transportation, did it.

It is important to note, in viewing the future as compared with the past, that the United States rose to its pre-eminent industrial position and its high standard of living on cheap energy minerals. It took vast amounts of energy to conquer the frontier and do all the construction needed to convert a raw wilderness into the world's most affluent and largest industrial society. Even as late as the 1940 to the 1960s period, much of that time the United States enjoyed $3-a-barrel oil, natural gas at around 15 cents a thousand cubic feet, and coal at about $4 a ton. Those energy economics were exceedingly helpful to a young and rapidly growing nation; but they will never return.

U.S. power may have peaked in 1945, when it used the ultimate energy weapon, the atomic bomb, to end World War II. At that time, the United States was the sole possessor of this fearsome form of energy; and it was the possession of a particular metal, uranium, which allowed the United States to arrive at this pinnacle of world power.

After explosion of the first two atomic bombs, the United States and the world may have entered into a new era, when it may be that wars of the future will not be fought by the violent methods of past wars. Possession of the atom bomb by a number of countries today and the probable destructive ramifications of atomic warfare, for all sides, make it unlikely that anyone could really win. Therefore, the military might of the United States may not be so important in the future as in the past. The new battlefield may be on the

127

economic and industrial fronts, drawing upon mineral and energy mineral resources chiefly for non-military purposes. In fact, it appears that the United States, even now, is in that conflict with its former military adversary, Japan, which is doing very well, indeed, on the new battlefield. But without access to energy and mineral resources, Japan would be almost totally crippled. This matter is pursued in a later chapter.

The United States rose to the top of the economic heap in record time; but in the process, it used up many of its own resources. The high grade ores of the Mesabi Iron Range are now gone, and all but one copper mine have closed in Upper Michigan. The United States is searching for oil off the frozen north coast of Alaska and in the deep waters of the Gulf of Mexico; and the USA is no longer, nor will it ever be again, self-sufficient in oil. Indeed, its oil reserves, once the largest known in the world, are now dwarfed by those of several other countries. And the United States has gone from being an exporter of energy and mineral resources to a net importer, on an increasingly large scale. In the process, the United States has also gone, in less than 20 years, from being the world's biggest creditor nation, to being the world's biggest debtor nation. At nearly $50 billion, oil imports are now a significant part of the annual balance of trade deficit; and these costs can only go up in the future. Is the luck of the USA also beginning to run out, along with its mineral resources? At the start of 1970, the United States was still self-sufficient in oil. In 1990, 50 percent of its oil was imported.

But in any case, the saga of the United States' astonishing rise in affluence and power will never again be repeated, anywhere in the world, for there are no more virgin continents to invade. The growth of the United States has truly been a phenomenon beyond compare. The question now is where does it go from here? Later we shall speculate on that.

And now to the story of Saudi Arabia, a country, largely desert, with a relatively small population: When the discovery oil well was completed on March 3, 1938, through 1979, it had produced more than 27 million barrels of oil and is still producing today. Saudi Arabia then was a nation of about three million persons, occupying an area about a third that of the 48-adjacent United States. Saudi Arabia had been a rather loose organization of tribes, many of which were desert nomads, together with some fishermen along the Gulf Coast. However, a very able desert chieftain, Ibn Saud, welded these units together into a kingdom, which was recognized by the British through a treaty in 1927. Ibn Saud became King of the Hijaz and Najd and its Dependencies. The country was renamed the Kingdom of Saudi Arabia, on September 22, 1932.

Oil, as noted, was discovered in 1938, little more than fifty years ago. But since that time, Saudi Arabia has leaped from being an insignificant, backward, third-world country, to becoming an economic giant. It has constructed, in the process, the world's largest and most expensive airport, the world's most modern communication system, and it has its own satellites. It has built railroads, a university, advanced medical facilities, miles of modern highways, and it owns its own fairly sizable airline — all this in 50 years' time. Actually, the early oil production was not very large, and therefore economic development proceeded rather slowly. Also, the years of World War II, 1939-1945, intervened to delay developments. As late as 1954, Saudi Arabia had only 147 miles of paved roads; but by 1986, only 32 years later, it had more than 50,000 miles of pavement.

The number of vehicles using these roads increased from only 60,000, as late as 1970, to nearly two million, in 1980. And the Saudi Public Transport Company now has 900 buses, providing low-cost transportation between all sizable cities and many villages.

In effect, Saudi Arabia has come almost as far in 50 years, in its standard of living and the use of modern facilities and equipment, as the United States did in 200 years. In terms of coming into the modern world, the Saudis arrived almost overnight — making the trip from a simple agricultural society to the twentieth century far more rapidly than did the United States. Oil did it.

It is truly said that money cannot buy happiness, but it can buy almost everything else, and that is the story of Saudi Arabia. It was able to make the leap into the twentieth century and obtain all the material things which characterize this era, by simply buying it.

The Saudis did not have to wait for the development of the telegraph, the telephone, automobiles, the electric light, the radio, antibiotics, television, and jet airplanes. All these had been invented and were already on the shelf, waiting to be purchased. The United States contributed much ingenuity and many inventions to the Industrial and Technological Revolution. Saudi Arabia simply bought it all, to a large extent with money from the United States — which paid for Saudi oil.

Saudi Arabia, along with its great oil deposits, also had another advantage which allowed it to come so far and so fast. It had a small population. Saudi Arabia has the world's largest oil reserve, the production profits from which are spread over relatively few people. As a result, raising the standard of living in terms of material things has been a comparatively easy task. It could not have been done if the oil wealth had been spread over, say, 100 million people. But the oil income has been substantial, and it was able to make a big difference in the lives of all its citizens.

And there is still no lack of Saudi oil to be produced, and it will be available in quantity, well into the twenty-first century. But there are some ups and downs, even for the Saudis with their great oil wealth. In the middle of the 1980s an oil surplus developed (temporary to be sure). The Saudis chose to support the price of oil by cutting back on production, to only about four million barrels of oil a day, far below the potential flow, and the 10 million barrels a day they once briefly produced.

But this caused a problem in that in bringing Saudi Arabia to its present affluent position, a number of projects were started, some of which are long term and are still in progress. Also, social programs of various kinds were implemented with regard to health care, education, and other matters. Saudi Arabia has been called by some observers "the world's largest welfare state." (The problem, of course, is how to maintain it indefinitely.)

The result was that the Saudi government found itself a little short of funds, because of its oil production cut-back and lower prices. They subsequently announced on December 30, 1987 that they planned to borrow up to $8 billion to help finance the 1988 budget. And to help cut the deficit, the 1988 budget was almost 17 percent less than the previous year's program. The new budget projected expenditures of $45.3 billion and income of $28.1 billion.

Thus, Saudi Arabia has indeed caught up with the United States, by entering into the era of government deficit financing. Welcome to the modern world!

But a final note to the good luck of Saudi Arabia might be in order, for oil is a finite resource. There is a Saudi Arabian saying: "My father rode a camel, I drive a car, my son rides in a jet airplane — his son will ride a camel." With some 250 billion barrels of oil reserves, it may be another generation or two before the Saudis are back to camels; but someday, indeed, the oil will run out and the present scene may well be looked back upon as the Golden Age.

Chapter 10

The Petroleum Interval

As Col. Drake watched oil slowly oozing from his 59-foot deep well in Titusville, Pennsylvania, not in his wildest dreams, nor in the vision of anyone else at the time, was it predicted that this black liquid would be the basis for huge industries not yet even thought of — the automobile, and aircraft industries ("what is an automobile or an airplane?"). If you had been there at the time, would you have visualized that this gooey black stuff could provide material with which millions of miles of roads around the world would be paved and that an entirely new way of life would emerge for the world, through the use of this substance?

If you had told the people gathered there that this dark liquid would eventually propel millions of people in cigar-shaped containers, with metal wings, around the United States and around the world, at an altitude of 35,000 feet and at a speed of 600 miles an hour, you would have been regarded as totally insane. Or if you had suggested that thousands of products, including clothing, medicines, insecticides and plastics ("what is plastic") would be made from that dark liquid, you would have been considered equally mad.

Or had you said that some of the then largely nomadic tribes in remote desert regions of the Arabian Peninsula had vast quantities of this material beneath their desert sands; and because of this, before the end of the twentieth century, they would become a nation that would be able to greatly influence the economies and futures of almost all the nations on Earth, any listeners would have scoffed. And if you had said that this circumstance would result in the greatest transfer of wealth the world had ever seen, and ultimately create chaos in parts of the international banking system, who would have believed you? How could this dark, thick liquid, oozing from the ground at Col. Drake's feet, cause all that! Your ideas would have been regarded as totally wild and unbelievable. Yet these are only some of the things that oil has done to and for the world, since it was discovered and began to be used in quantity.

It is difficult to overstate the changes which have taken place: i.e., starting in 1859 when world oil production was only a few barrels a day, to the current 50 million barrels a day, or more. No other material in the world has so profoundly and universally changed the world, in such a short time, as has the energy mineral oil. Iron might be the nearest candidate for the claim but the influence and use of iron spread out over the world during many centuries, whereas in less than 100 years oil accomplished its effects and changes — almost instantaneously in the perspective of human history.

It may also be difficult to overestimate the changes which will take place in the world as petroleum supplies gradually diminish toward the point of exhaustion. It can be observed that the changes which oil brought were, for the most part, pleasant ones. Its sudden arrival on the scene has been a circumstance which has done many good things, for many people, including, you, the reader. However, the decline in oil supplies may have the opposite and somewhat unsettling effect, Fortunately, that phase of the Petroleum Interval will be gradual, so there may be time to make adjustments.

The Stone Age lasted for hundreds of thousands of years, perhaps more than a million. The Copper Age was shorter, but still many centuries long; the Bronze Age was a bit shorter, but still several hundred years; and the Iron Age has been with us for many centuries.

The Petroleum Interval began on August 27, 1859, with Col. Drake's discovery that oil could be obtained by drilling. It has now lasted slightly more than 100 years. The term "interval" has been chosen, rather than "age," for the use of petroleum, in great quantities, as we are doing now, will be a much shorter period than the "ages" of the past, just mentioned. It will be but a brief, bright blip on the screen of human history, lasting in significant form for perhaps about 300 years. How fortunate we are to live in it — that is, those of us who are able to enjoy its benefits. There are many people living today who will never get into the Petroleum Interval, in any significant fashion. There are more than a billion Chinese, most of whom will only get a passing, distant glimpse of this time. The same is true for more than half a billion inhabitants of India, and many people in Africa and South America. The rise of petroleum production and the many and varied uses discovered for this most useful energy mineral is truly a singular event in the history of the Earth's resources and their exploitation. Petroleum is also doubly valuable because it can be used as the raw material for many products — a service which some other forms of energy (e.g., uranium, sunlight, wind, and electricity) cannot provide.

It is worth repeating that no other natural resource ever produced in the past has had such a rapid and widespread impact on the world as petroleum. In this chapter, we shall explore this singular event, in some detail (For the sake of clarity, we note again, as in the Foreword, that the term "petroleum" includes both oil and natural gas; but in this chapter, we discuss chiefly oil).

Oil is by far the most convenient to handle and transport, high-energy-per-unit resource we have. It can be pumped, carried in cans, put in fuel tanks to power mobile machinery, stored for long periods of time; and it burns with little or no ash that needs disposal. It can be hauled great distances, indeed around the world if need be, in huge ocean-going vessels, and it can be loaded or unloaded easily and quickly by pipes and pumps. There is no other energy form, as versatile in its uses, and which can be transported so far so easily, and, except for uranium, deliver such a large net energy return, at the far end of the trip, as oil. And uranium is by no means so versatile in its end uses.

Coal can be transported great distances but the amount of coal it takes to transport a cargo of coal is greater in terms of the amount of energy used, compared with the amount received at the destination, than in the case of oil. Also, coal is dirty to handle, bulky, and leaves considerable ash when burned, as well as other pollutants, such as sulfur; and contributes much more to air pollution than does the burning of oil, although oil cannot be held harmless in this regard, as any resident of Los Angeles can testify. Still, oil is clearly the world's premier fuel.

The United States is where the modern oil industry was born, and the United States produced the initial group of oil-finders. And this is where petroleum geology and petroleum engineering studies received their start, and where a number of universities set up departments devoted specifically to these disciplines. The products of these schools, geologists, and petroleum engineers, ranged first over the United States; then more widely, to Mexico, Canada, and South America; on to the Middle East and North Africa; and finally all across the globe, from the arctic to Australia and the Indonesian Archipelago.

In 1920, two-thirds of the world's oil came from the wells of the United States. During the period from 1859 to 1939, 64 percent of all the world's oil production came from the United States. The United States had produced the bulk of the world's oil to that time and used that oil to help it reach the top of the world economic heap. But the importance of U.S. oil production, in terms of *world* production, was dropping.

From producing two-thirds of the world's oil in 1920, the United States, in 1988, was the source of only about 15 percent of world annual oil production. The center of oil production had moved, and become less concentrated. Many countries now produce substantial quantities of oil.

In the 1980s, the Soviet Union was the world's largest single oil producer, with the United States second. Saudi Arabia had the potential at that time of being the world's largest producer; but wanting to stabilize oil prices, the Saudis were only producing about four and one-half million barrels of oil a day. At that rate, their 167 billion barrels of oil reserves would last slightly more than 100 years. If one accepts the recently revised Saudi estimate of 252 billion barrels of oil reserves, their reserves would last somewhat more than 150 years. Either way, the Saudis have a long oil producing future ahead of them. At the same time, the United States was producing about eight million barrels a day, against a proven reserve of about 25 billion barrels, which means, without future discoveries, the reserves were equivalent to slightly more than eight years' production. Of course, in both cases, in the United States and in Saudi Arabia, additional oil discoveries will be made. However, the United States is the most thoroughly drilled up country in the world and is fast running out of acreage in which to make further major oil finds. The chances of finding any significant number of major oil fields, especially any the size of Prudhoe Bay, from which one-half the oil has been produced, are slight. The balance of power in terms of oil production has definitely left the United States, never to return. In contrast to the limited further potential for U.S. oil discoveries, the Saudis, already owners of the world's largest oil reserves, are continuing to extend their oil fields.

In 1989, about 190 kilometers south of Riyadh, and well outside the old Aramco concession area, Saudi Aramco drilled a well in the Al Hawtah region with a production potential of 8,000 barrels a day of high quality sweet (low sulfur) crude oil, from rocks of Paleozoic age, which lie beneath the producing strata of the original ARAMCO concession. This strike opens up large new territories and stratigraphic zones which the Saudis can presumably bring into production at some later date, expanding still further their already huge oil reserves.

It is worthwhile to examine in some detail who has the oil at the present time, and who is likely to have it in the future. We have come a long way in oil exploration from when Col. Drake simply drilled near some oil seeps. Oil occurs, not on mountain tops, but in basins in the Earth's crust, which slowly sank and into which organic materials were deposited. Most of these basins were in shallow areas of the ocean, along the continental margins or actually on the continents themselves, which have from time to time been invaded by the sea. Accordingly, to determine where the oil is, one simply looks for these basins around the world. So far, some 600 have been identified; and this number we now know, from rather thoroughly mapping the broad geological features of the world, is close to the total number of basins which exist. Of these, about 400 have been explored by the drill, to a greater or lesser extent. The other 200 are in fairly hostile environmental areas, such as in the Antarctic region, and have been explored only in a minor way, if at all.

In the approximately 400 basins which have been drilled, we can determine two basic figures: how much oil has been found, and in how many cubic miles of sediments this oil occurred. If you divide one figure into the other you get a world average of how much oil you find per cubic mile of sediment in these basins. The next step is to calculate the volume of sediments in all the oil basins, in cubic miles or whatever unit you choose; and then multiply that figure by the amount of oil which, on the average, is found per cubic mile or other unit. Thanks chiefly to the reflection and refraction seismic technologies, and advanced gravity and magnetic studies, enough information is available to estimate fairly accurately the volume of sediments in all the basins of the world, drilled and undrilled. Geophysical ships, helicopters, airplanes, trucks, and strong backs have hauled technical equipment to the near and far corners of the Earth; and the inventory of basin size and sediments is quite good. The world is no longer unexplored.

It has been measured and we know it quite well. From these calculations, and the presently known drilled resources, it has been estimated by Masters and others (1987) that as of 1985, world oil resources were *initially* about 1744 billion barrels, of which approximately one-third or 544 billion barrels have already been produced. As part of this worldwide assessment, these authors further stated that they did not believe any oil provinces with a potential of 20 billion barrels or more had been overlooked. (Note: 20 billion barrels would be about two Prudhoe Bay fields). Briefly, what they state is that we now have a pretty good grasp of world petroleum geology, and of the locations and volumes of the basins where oil has to occur.

Although there is apparently more oil yet to be produced than has been used so far, the important fact is that the areas which produced most of the oil previously are not necessarily the areas which will produce most of the future oil. Some regions moved into oil production much earlier than other areas, and the United States was the very first. Some places had less or more oil than other places, and production life and peaks will differ accordingly. Combining these factors means that some countries are just now beginning to be producers (for example, Yemen), and others are now in essentially peak production (for example, USSR), and some nations are past their peak (for example, the USA).

The United States is the world's largest consumer of oil. With six percent of the world's population, the U.S. has recently been consuming about 30 percent of the world's oil, while producing only about 15 percent of it. Thus the position of the United States, with regard to its domestic oil supply, is worth special examination, as its consumption is such a large factor in the world market; and indeed, oil is priced world-wide in terms of dollars.

Here are some basic facts regarding the history of U.S. oil production. From the time the Drake well was drilled in 1859, to the year 1909, oil production built up to about 500,000 barrels a day. This was, at that time, more oil than all the rest of the world *combined* was producing. The United States continued to produce half or more of the world's oil until as recently as the early 1950's. Oil production reached its peak in 1970, at about 9.3 million barrels a day, to which can be added the liquids which come from the production of natural gas, giving a total liquid production of about 11.3 million barrels a day.

This amount is what the United States also used in the year 1970. The fact that the United States reached the peak of its oil production at approximately the same time its domestic *consumption* equalled all the production is of great importance, for at that moment, the control of oil prices shifted to the Middle East. As long as the United States had surplus production beyond its own needs, even if foreign oil producers raised their prices, the United States could still produce all the oil it needed and price it as it wished. And with some to sell to world markets, the United States could influence world prices as well. But when there was no longer any surplus capacity and some oil had to be bought abroad, then the price of oil was no longer determined by the United States. By that single event, in 1970 the United States lost control — at least in part — of its economic destiny.

In 1930, the great East Texas oil field was discovered — the largest field ever found in the 48-adjacent states. There were no regulations then on oil well spacing for efficient production, or limits on rates of production to prevent wasteful reservoir engineering practices. The wells were allowed to run wide open and the price of oil got down as low as four cents a barrel! On an emergency basis, the Texas Railroad Commission was given authority to limit oil production of each well — an "allowable," it was called.

The Texas Railroad Commission retained this authority permanently; and each month, until the United States (in 1970) could no longer supply its own needs, the Texas Railroad Commission, and similar regulatory agencies set up in other oil producing states, issued an "allowable" for each well. By this method, production and price were controlled. Just enough oil was "allowed" to be produced to take care of the United States' needs, at a relatively stable price, and to supply what foreign markets we supplied at that time. Increase the amount of oil produced and the price would drop; decrease the amount and the price would rise; and as the United States (to the year 1950 or so) was still producing half or more of the oil in the world, this system, in effect, controlled world oil prices.

After about 1970, the "allowable" system was not used basically to control the *price* of oil, but to see to it that the wells were produced efficiently (running a well wide open is wasteful and does not allow the maximum amount of oil to be recovered from a reservoir).

This earlier operational "allowable" procedure gave stability and predictability to the oil industry. Under these conditions, companies could plan ahead, and raise capital from investors. The industry flourished, and it provided American consumers with the greatest quantity and cheapest and broadest spectrum of petroleum products of any nation in the world. There were no shortages. But as production declined and demand continued to rise, the wells were eventually allowed to run up to their MER, which is the maximum efficient rate of production, beyond which higher production would injure the well and producing zone (resulting in recovery of less oil, in total, than otherwise).

But late in 1970, even with the wells running open to their MER, the oil needs of the United States were not completely satisfied. At that point, the United States became increasingly and *permanently* dependent on foreign oil and lost control of the price of oil. Saudi Arabia became the "swing producer," meaning it had "shut in" production by means of which it could increase or decrease the world oil supply and therefore influence, if not control, its price. This marked a milestone, for from then on the United States would never again control oil prices. It lost an economic weapon and gave it to another country.

What is the situation today? The United States is the most thoroughly oil-explored and drilled nation in the world. Of the somewhat more than three million wells drilled in the world, more than two million have been drilled in the United States. Except for certain offshore areas, chiefly in the deeper waters of the Gulf of

Mexico, and in arctic and sub-arctic waters, and a portion of the Alaska North Slope, which includes a national wildlife refuge, there is hardly an undrilled area large enough that a major oil field could be discovered.

A major oil field covers a number of square miles. There are just not many reasonably prospective areas where there is space between dry holes, producing wells, or depleted wells large enough to accommodate a major oil field. In contrast, prospects are a great deal better abroad, which is the reason, in 1988, that ARCO, which has in the past depended heavily on domestic oil reserves and was the major company in the Prudhoe Bay field, spent 60 percent of its exploration budget overseas. Texaco spent half of its exploration budget that year outside the United States. Other companies are similarly finding that the "oil patch abroad" is better hunting ground than is the USA.

Perhaps in the United States one could simply drill deeper to get more oil in already proven areas? But there are limitations here also. Kansas, for example, has a long history of shallow oil production. It has historically been a great place where a small operator, with limited capital, could explore for oil inasmuch as the wells are almost all less than 6,000 feet deep. "Post-hole" drilling, this is sometimes called. The answer is that, in most of Kansas, 10,000 feet down (or less in places), you would be in granite, which is notably not an oil-producing rock.

In other regions of the United States, where sedimentary rocks, the kind in which oil occurs, are very thick, there is the matter of the geothermal gradient. That is, as you go down into the Earth, it gets hotter, to the extent that generally below 15,000 feet (with some exceptions) oil cannot survive, and only gas exists.

There is a further problem in drilling deeper for oil. The deeper one goes, the greater the pressure from the overlying rock, the pore space available to hold the oil is reduced, and the individual pores tend to be smaller. So deeper wells are generally not such good producers as are the wells of more moderate depth, although, again, exceptions occur. Thus, in going deeper, one may incur the higher costs of deeper drilling but expect, perhaps, less production than from shallower wells.

Also, in terms of pore space, comparing the United States with some of the other major producing areas of the world, such as Mexico, and particularly Saudi Arabia, much of the oil in the U.S. is obtained from sandstone reservoirs. These tend not to have a great deal of pore space in which to hold the oil; and also the ability of oil to flow through, and from, these reservoirs is limited in many cases, especially if there is a lot of cementing material (clay or calcite, for example) in the sandstone. In contrast, some of the Mexican wells and many of the Mid-East wells are producing from limestone, which commonly has relatively large pore spaces, being in some places almost cavernous.

Without burdening the reader with technicalities, one other comparison is instructive. Permeability — the ability of a rock to transmit a fluid such as oil — is measured in terms of a unit called the 'Darcy. In the case of Saudi Arabia, some formations there have permeabilities which are measured in the range of a full 'Darcy or more. In the United States, permeabilities are commonly found to be in a range which is measured in millidarcies — that is, thousandths of a 'Darcy. Partly as a result of this difference in permeabilities, the average production per-well per-day in the United States is less than 14 barrels, whereas the average daily production per well in some of the major Arabian oil fields is in excess of 10,000 barrels, and in some cases exceeds 15,000 barrels a day. This high rate of production could only exist with good permeability in the producing formations. Good permeability also means that a given well can more efficiently drain a larger area than can a well which is producing from a "tight" formation — one with low permeability; and therefore fewer wells have to be drilled; thus reducing the cost per barrel. The Mid-East reservoirs and many other reservoirs abroad are just much better reservoirs than what are now left in the United States.

Mexico apparently holds the world record for daily oil production from a single well. In 1910, the famous Potrero del Llano well Number 4, of the Mexican Eagle Oil Company was brought in with an initial flow of 100,000 barrels a day. But it soon got out of control and flowed about two million barrels, most of which was lost until the well could again be brought under control. That single well ultimately produced more than 100 million barrels of oil. No such well, or one even close to it, has ever been drilled in the United States. In 1985, the United States needed about 650,000 wells to produce about nine million barrels of oil per day. At the same time, five OPEC countries in the Middle East also produced about nine million barrels of oil per day and did it from only about 3,000 wells.

As the reasonably accessible prospective oil areas are drilled up in the United States, oil companies have been forced to go into what they term "hostile" regions, if they want to continue exploration in the USA. Such areas include Norton Sound, off southwestern Alaska, a region of severe storms, and difficult working conditions (and so far proving rather disappointing in drilling results). The cost of drilling in these areas is very high. The most expensive oil well ever drilled in the world was drilled by Sohio Oil Company, their Mukluk Number 1, on an artificial island off the Alaska north coast. It cost $1.6 billion in total, and was a dry hole. The structure which was drilled was gigantic, but the oil had escaped the trap, some time in the geologic past.

To those unfamiliar with the oil industry, the cost of finding oil might simply seem to be the cost of drilling the well. However, that may be only a small part of total expenses. There is the cost of acquiring the leases (sometimes paying a competitive bonus for them even though there is no assurance that there is any oil to be found therein; the federal government and state governments get money from lease sales of their lands, and they get the lease money regardless of drilling results. If any oil is found, they also get a royalty from each barrel produced. They cannot lose, and, of course, this cost is passed on to the consumer). Then the area must be explored, to determine the exact drilling site.

This exploration is done not only by geological studies but also by the use of expensive geophysical equipment designed to "see" into the Earth and find hidden (and what are hoped to be) oil-bearing strata and structures. There is the cost of processing and analysis of all the geophysical data (a huge task, involving billions of computations, which can only be handled by computers). Then the data have to be interpreted by well-trained people. There is the cost of laboratory analyses of rocks, for their organic content, to see if they might really be source beds for oil.

If drilling is conducted in the ocean, one must first study the ocean floor and what lies beneath it, by means of a ship, fully manned with highly trained technical personnel and with a great deal of expensive equipment. Finally, there is the cost of all the equipment, which goes way back to the cost of mining and processing the ore that ultimately went into producing the steel tools with which to explore for oil, steel for making the drilling derrick and also provide the steel casing for the completion of the well — in some instances two to four miles of steel casing.

All these costs can be thought of and calculated in terms of energy costs. There is the energy cost also of building the ship and the energy cost of the materials used to build the ship, if a ship is to be involved in oil exploration. And there is the energy cost of running the ship, or the energy it takes to haul the geophysical equipment over the tundra of the arctic regions, or through tropical jungles, or perhaps through the air to remote sites by helicopter. There are the energy costs of doing the basic geological field work, if the prospect is on land. The gasoline to run the Jeep that I used in Peru, and in western United States, as a petroleum geologist, had to eventually be accounted for by getting it out of the oil wells which this activity helped to discovery. If this work did not result in oil discoveries, then these energy costs have to be charged by the oil industry against wells somewhere else which do produce.

All of this energy, expended in thousands of ways used to ultimately discover oil and produce it, has to be added up and compared with the amount of energy in the oil which such efforts produce. This ratio of energy used to net energy recovered is the all-important profit/energy ratio. As we have to drill deeper to find the same amount of oil, and move into more difficult areas in which to operate, the ratio of profit to energy declines. Already, in some situations, the oil found is not equal to the total energy expended. Also, although some wells flow initially, all wells eventually have to be pumped. Pumping oil is expensive, particularly if it is being pumped from a considerable depth, for it takes energy to move the steel pumping rods up and down — in some cases, as much as three miles of these rods.

The rods also wear out from the sand or other abrasive materials which may come into the well or be "eaten" out by the sulfur compounds and other corrosive chemicals which may be in the oil. The rods then have to be replaced, at great expense. Finally, most wells eventually "go to water," so it may become necessary to pump several barrels of water, just to recover one barrel of oil. The water is heavier than oil and costs more to pump than does oil; and also, it has to be disposed of in some fashion. It is expensive to operate an oil field.

Ultimately, all wells have to be abandoned when the energy recovered does not equal the energy expended. This abandonment occurs even when there is considerable oil still left in the ground. There are two figures involved here. The amount of "initial oil in place" and the amount of "economically recoverable oil." The latter figure can never equal the former. In general (at least in the past), more than half of the oil in place is never recovered.

The most significant fact in the U.S. oil industry trends has been the decline in the amount of energy recovered compared to the amount of energy expended. In 1916, the ratio was about 28 to 1, a very handsome return. By 1985, the ratio had dropped to 2 to 1, and is still dropping. The Complex Research Center at the University of New Hampshire made a study of this trend. The conclusion was that by the year 2005, at the latest, it will take more energy (on the average) in the United States to explore for, drill for, and produce oil from the wells than the wells will produce in energy.

As already noted, the United States peaked in oil production in 1970 and has declined since then. This decline is likely to continue for the basic reason already cited — lack of new, large areas in which to explore. As has been flippantly but quite factually stated, "Exxon ran out of real estate." To further complicate the matter, the easy oil has been found. This includes shallow oil found, in some cases, by persons with little or no knowledge of geology, who simply went around drilling holes in the ground. Also, earlier much of the oil was found in structures called anticlines, which are simply up-folds in the Earth's crust. Some of these are clearly visible on the surface and it takes only the most elementary geological training to recognize them. These have been drilled.

Now the search for oil in the United States has turned to oil traps not visible on the Earth's surface. These include buried anticlines, buried sand bars, old shore lines, and stream channels. These oil traps may be buried at depths of many thousands of feet. Finding them involves costly exploration and considerable drilling. In this regard, there is a useful ratio of oil found per foot drilled. In the United States, during the early 1930s, about 250 barrels of recoverable oil were found per foot drilled. By the 1950s this figure had decreased to about 40 barrels a foot; and by 1981, it was down to 6.9 barrels per foot drilled. To exacerbate the matter, the cost of each foot drilled was going up, for the wells were getting deeper and deeper; and therefore the footage was getting more expensive.

When oil prices tumbled in the middle and late 1980s, the United States' oil-finding community, faced with the highest exploration and production costs in the world, drastically cut back. The number of operating drilling rigs dropped from more than 4,000 to less than 800. Many operators went out of business. Other oil people decided to go "drilling on Wall Street." That is, they found it was cheaper to "buy oil already discovered," by buying oil companies, than it was to go out and drill for it. There were corporate raiders such as T. Boone Pickens searching through the industry, the result of which was a "restructuring" of the industry by mergers and acquisitions. But none of this activity found a barrel of oil. It simply made lawyers, accountants, and some entrepreneurs some quick money; and the American oil industry, in general, came out of all of this considerably poorer and weaker. It will never be the industry it was in the 1970s and before.

Another disturbing trend in U.S. oil exploration is that the discovery rate of new oil fields is quite disappointing. From 1977 to

1986, less than two billion barrels of recoverable oil were discovered in new fields. The additional oil reserves that were reported (which totaled about 20 billion barrels) were found around the edges of already known fields, and by drilling inside known fields. Clearly, oil hunting has become much more marginal than in the past; and only a few remote areas may have the potential of large discoveries. Most notable of these is a region on the Alaska North Slope, east of Prudhoe Bay, which includes the controversial Arctic National Wildlife Refuge.

What is happening *in general* to the United States' oil situation is well illustrated by Texas, which has long been the major oil producer among the states. In 1960, Texas had oil reserves of about 14.8 billion barrels. By 1975, that figure had dropped to about 10 billion barrels, and by 1985 the figure was only about 7.8 billion barrels, hardly more than half the reserve figure of 1960. In 1960, each Texas oil well (on the average) produced about 4630 barrels of oil for the year; and in 1985, each well annually produced about 3950 barrels. Obviously in the great oil producing state of Texas, more oil is being produced than is being found.

In the case of natural gas in Texas, the trend is much the same. In 1960, Texas had reserves of about 119 trillion cubic feet. By 1975, this figure was down to 71 trillion cubic feet; and in 1985, the figure was 40 trillion cubic feet. Overall, the petroleum reserve figures for the United States have shown a steady decline the past twenty years or more, with the exception of the banner year when Prudhoe Bay was discovered. There is little likelihood that this trend can be reversed for very long. A major discovery in the area east of Prudhoe Bay in, or in the vicinity of, the Arctic National Wildlife Refuge would do it again; but only temporarily, for the estimated amount of oil which most probably could be recovered from this area, according to both industry and the U.S. Geological Survey estimates, is about 3 1/4 billion barrels or only approximately one-third of the original amount of recoverable oil in the Prudhoe Bay field.

It is worthwhile, however, to take note of a study headed by Dr. William L. Fisher, director of the Texas Bureau of Economic Geology, made public in 1989, which comes up with the rather amazing figure of 247 billion barrels of U.S. remaining oil reserve. This figure is reached by assuming both the use of existing technology and, to some extent, that yet to be devised and an ultimate price for the final barrels of $50. As we have already noted, most oil fields at present recover less than half of the original oil in place, so there is no doubt that more is there to be recovered. How *much* can be recovered is the multi-billion dollar question. At a price and with a time lag for the putting in place of existing technology and the development of new recovery techniques, more oil no doubt can be obtained; but it will be costly. This statement also applies to all oil fields around the world. But we return to the figure of about 26 billion barrels of conventional oil reserves for the United States, when we compare the U.S. position with other nations of the world, using the current standards of measuring oil reserves.

Whereas oil, chiefly in the form of gasoline, enters into the life of nearly every American every day, it is amazing how little the general public knows about the basic facts of oil consumption and production in the United States.

In the late 1980s the United States was using about 17 million barrels of oil a day, and producing about 7.8 million barrels of oil and about 1.6 million barrels of liquids, condensed from natural gas. The giant Prudhoe Bay oil field had initial reserves of about 9 1/2 billion barrels of oil. How long would that amount last the United States, if the USA used only that source for its entire supply? The answer is about 1 1/2 years! But such giant oil fields are rarely found. The Prudhoe Bay field is the second largest field ever found in the United States and none has been discovered since the Prudhoe Bay find in 1969. If the Arctic National Wildlife Refuge and adjacent north slope areas east of Prudhoe Bay are opened for exploration, the amount of predicted oil to be found — 3 1/4 billion barrels — is equal to just about one-half year's consumption in the United States. We could cut this use considerably if we wanted to reduce our numbers of cars, drive less, build smaller houses, and employ other conservation measures. But so far, we are apparently quite unwilling to give up any part of our "oil standard of living" and indeed it is very much oil-based.

But as we watch our conventional United States' oil reserves decline, we are clearly living off of our capital. Some of the most promising prospects that remained in recent years have been disappointing. On the East Coast, the Baltimore Canyon area was regarded as an excellent prospect; but the Shell Oil Company drilled four wells in what at that time was the deepest water in which any wells had ever been drilled — 7,000 feet — and all four wells, very expensive wells — were dry.

The largest undrilled structure in the eastern Gulf of Mexico, the Destin Dome, came up a duster. About 600 miles offshore of southwest Alaska, in very rough seas, is the Narvin Basin, as big as the Gulf of Mexico. Amoco, ARCO, and Exxon, with the aid of the most sophisticated state of the art geophysical equipment to help them locate the best drill sites, drilled nine holes, all dry.

World-wide we also passed a milestone in 1986, for that was the first time that global production of crude oil and natural gas liquids was greater than the amount discovered. Big oil fields are getting harder to find.

Since serious oil exploration began in 1859, over 40,000 oil fields have been found in the world. The great majority of them, however, have little impact on world oil production. There are only 37 so-called "super-giant" oil fields (those with more than five billion barrels of recoverable oil). But these 37 fields originally held more than half of the oil discovered in the world so far. Significantly, 26 of these super-giants are in the Persian Gulf area. Libya, the USSR, Mexico, and the United States each have two, China, and Venezuela have one each. The increasing concentration of known world oil reserves in the Middle East is of utmost importance for the future.

The trend in Soviet production, wherein they are currently peaking out, almost ensures that the Middle East will continue to be a region of confrontation and international political intrigue. Currently, the world's largest oil producer at about 12 million barrels a day, the Soviets also, like the United States, are finding that increased efforts are yielding less and less oil.

In what is now the most important oil province in the USSR, the Western Siberian Basin, the Russians report that in 1970 it was necessary to drill about 865,000 feet of hole to get an increase in production of 20,000 barrels a day. In 1980, they had to drill 2.84 million feet to add 20,000 barrels of oil per day. By 1985, the footage drilling figure had reached 7,000,000, to add that 20,000 barrels a day. Also, new wells being drilled were obviously located in less productive areas, for in 1976, the average new well initially yielded about 730 barrels a day. In 1985, the average yield of each new well was only about 300 barrels a day. The Soviets, like the United States, are also clearly running out of the more prospective acreage. In addition, in 1985 General Secretary Gorbachev reported that the costs of producing an additional ton of oil (about six barrels) had risen 70 percent in the past 10 years. Moreover, the USSR has had to move northward into more hostile areas (thicker permafrost, colder weather, more remote locations, more difficult and longer road-building problems) increasing the difficulties of conducting oil exploration.

There is, as we have just seen, an overwhelming amount of data concerning the basics of the Petroleum Interval we now enjoy. We just covered some of these facts, but perhaps the most important and significant figures, which show who now has the oil and how much. To establish reasonable credibility for these figures, it should be stated that we do have a fairly accurate idea of the number of oil basins in total, both drilled and undrilled, and the total volumes of sediments they contain.

We first examine figures for the amount of oil in the form of proved reserves, which is oil already discovered and the volumes established by drilling, and at least some production history. The second set of figures presents the total estimated recoverable oil reserves, both discovered and undiscovered — the latter category based on the concept we have already described, by estimating the undiscovered oil from the volumes of sediments in the undrilled basins or portions thereof, and using a world-wide average of oil recovered per cubic mile of sediment.

As of the start of 1989, estimated world proved oil reserves (oil found, and which can be produced by current technology at market prices) stood at 907 billion barrels. Of this amount, the oil is distributed among the principal producing countries as follows: (Source: Oil and Gas Journal, December 26, 1988, p. 46-47).

Estimated Proved Oil Reserves
(In billions of barrels, as of 1/1/89)

Saudi Arabia	169.9*
Iraq	100.0
Iran	92.8
Abu Dhabi	92.2
Kuwait	91.9
USSR	58.5
Venezuela	58.1
Mexico	54.1
United States	26.5
China (mainland)	23.3
Libya	22.0
Nigeria	16.0
Norway	10.4
Algeria	8.4
Indonesia	8.2
Canada	6.8
India	6.4
United Kingdom	5.2
Egypt	4.3
Dubai	4.0
Brazil	2.5
Argentina	2.3
Angola-Cabinda	2.0
Ecuador	1.4

*Saudi Arabia claims 252.4 billion barrels as of January 1989. If correct, this would be more than one-fourth of the world total. We have already noted that Kuwait, in its 7,000 square miles, has more than three times the proven oil reserves of the United States, in its 3½ million square miles. Abu Dhabi, in 25,000 square miles, about the size of West Virginia, has nearly four times the oil reserves than in all of the 50 United States combined.

However, these proven oil reserve figures many not be as significant as the figures for total estimated oil reserves. These are what may (from now to the end of production) be produced by each country. Subtracting the oil already produced, the remaining conventional oil reserves, which include both discovered and *estimated undiscovered oil*, which can be recovered by present technology and at reasonable market prices, are estimated as follows:

Estimated
Total Recoverable Conventional Oil Reserves
(Figures in billions of barrels, from Riva, 1987)

Saudi Arabia	208.6*
USSR	158.0
Kuwait	97.2
Iraq	85.7*
United States	84.4*
Iran	79.1*
China	58.6
Mexico	55.0
Venezuela and Trinidad	54.7
United Arab Emirates	50.7*
Libya	31.9
Norway	27.0
United Kingdom	20.0
Indonesia	17.0
Algeria	13.0
India	9.0
Malaysia and Brunei	9.0
Egypt	7.0
Brazil	6.0
Australia and New Zealand	6.0
Ecuador	4.0
Argentina	4.0
Colombia	4.0
Peru	3.9
Tunisia	3.8

*Note that these figures do not correlate well with those tabulated for current estimated total proven reserves. For example, Iraq now claims 100 billion barrels of proved reserves, which is substantially larger than the figure given in this table of what might be expected in total. If Saudi Arabia's most recent figure of 252 billion barrels of proven reserves is accepted, their ultimate production capability would probably be 300 billion barrels or more. Abu Dhabi (of the United Arab Emirates) has issued a huge estimate of reserves, which is not reflected in the immediately foregoing table. The estimate recently issued by Dr. William Fisher and his group of 247 billion barrels remaining discovered and undiscovered oil reserves in the United States, is also not included. It is difficult to assess the validity of all of these estimates, as there are a number of variables, including projections of what future technologies might do. In regard to foreign oil reserve estimates, in the case of OPEC, their producing quotas are based, to some extent, on proved reserves and it tends to be in the respective governments' interests to report the largest reserves possible in order to obtain the greatest possible OPEC production quota. In the case of other governments, foreign loans are based, in some cases, on the oil reserves the country is presumed to have. (This was the case of Mexico in the past). The figures which have been quoted here, in the foregoing tables, are the best available, reasonably firm estimates of oil reserves, as of this moment, subject to the reservations and qualifications which have also been expressed.

Looking at the situation by region, the following tabulation again shows how geographically concentrated oil reserves are. These are remaining recoverable oil reserves, both discovered and undiscovered (same figures as in previous tabulation but grouped by region). Note that these figures add up to more than 100 percent, because the regions overlap (From Riva, 1987).

Region	Percent of World Total
Eastern Hemisphere	79*
OPEC	56*
Persian Gulf	44*
Non-OPEC	23
Saudi Arabia	17*
USSR	13
United States	7*
China	5
Mexico	5

*These figures are subject to the variables already discussed. Accepting the variables would basically mean that the dominance of the OPEC nations and the Persian Gulf region for future oil supplies is probably increased from these statistics. The 247 billion barrel figure for the United States probably cannot be realistically compared with the other figures which have been cited, as the bases for the figures are not the same.

If the bases used for inferred reserves of 247 billion barrels for the USA were applied to the Persian Gulf area, the dominance of the Gulf area would be increased substantially and probably would be appreciably more than three-quarters of the world oil reserves.

It is especially significant from the United States' point of view that approximately four-fifths of the remaining conventional oil reserves of the world are in the Eastern Hemisphere. Important also is the fact that the Middle East countries are the world's lowest cost oil producers, with an average cost of less than $2 a barrel.

Saudi Arabia probably has a cost of 50 cents or less. To replace the oil being produced in the United States, the cost is now somewhere between $9 and $22 a barrel, depending on where the drilling is done. Oil production in countries outside the Middle East is costly, particularly in the harsh environment of the North Sea, where Norway and Great Britain have nearly all their production, and in offshore Alaska and the Canadian arctic.

Saudi Arabia and other Persian Gulf nations also have very large fields, which can produce for a long time, so that additional, vigorous exploration is not really needed for some years to come. They can live very well on the oil already found. This is not true of the United States, where we are using more oil than we can produce; and many of the fields are small, with the average well producing less than 14 barrels a day. This compares, as we have noted before, with the daily well production in some of the fields of Saudi Arabia of from 10,000 to 15,000 barrels a day.

Saudi Arabia is currently producing only about four million barrels a day, to stabilize prices, but can easily produce 10 million barrels a day or more, if it wished. The United States has no reserve production capacity. The wells are running as wide open as they can be efficiently run. This is also true of the Soviet oil fields. Thus, it seems clear that oil production more and more will center in the Middle East. Allah will allow the Arabs to have the last word on oil. OPEC will probably survive and eventually emerge with fewer but stronger members. Those OPEC nations outside the Middle East will gradually drop by the wayside, as their oil runs out.

As we have noted, some countries have produced their oil much earlier and faster than other areas. Some have passed their peak, some are about at their peak, and some are still coming up to their full potential. It is important to observe, also, that oil is now being produced by various countries, at markedly different rates relative to their reserves. Simply stated, some are going through their reserves a lot faster than are others. This is the so-called reserve to production ratio, written R/P. The United States, for example, has already probably produced about two-thirds of what was its ultimate amount of conventionally recoverable oil, and has a current R/P of 8/1. The Soviet Union has produced about one-third of its ultimate oil resources, and has an R/P of 13/1.

In marked contrast stands Kuwait, with an R/P of 158/1, and Saudi Arabia, currently with an R/P of 114/1. (Or if the recently revised figure of 252 billion barrels of reserve is correct, then the R/P is about 172/1). These two Gulf nations have long oil-producing futures ahead of them, even at substantially increased rates of oil extraction; and in such circumstances, will have an important influence on world economies for many years to come.

The United Kingdom, very much a Johnny-come-lately on the world oil scene, is also going to disappear fairly fast, as they have an R/P of 9/1 and no great area left to explore — estimated total remaining reserves, both discovered and undiscovered, being about 20 billion barrels. Iran, chronic trouble-maker in the Persian Gulf region, for better or for worse can look ahead to many years of oil production at current rates, with an R/P of 74/1. Looking at these and other figures: At 1989 production rates, the United States has already begun to decline; the United Kingdom will probably start to decline in the early 1990s; Canada's conventional oil production will probably begin to decline in the mid-1990s, and it will no longer be self-sufficient in oil. The Soviet's oil production is now about at its peak (12.3 million barrels/day). They perhaps can sustain this rate for a few years, although a slight decline actually occurred in 1988. The near approach of the permanent decline in Soviet oil production is a significant matter to consider, in regard to future events in the Middle East. It is now evident that one of the main reasons for the Soviet Union's abortive attempt to take control of Afghanistan, in the 1980s, was to extend the USSR's influence closer to the oil-rich Persian Gulf.

Nigeria can sustain its current production to about the year 2026, Mexico to the year 2036, Libya to 2061, Norway to 2066, Saudi Arabia to about 2103, Iran to 2096, and Kuwait to the year 2175. Qatar, in its 11,437 square miles, and somewhat less than 300,000 people, sits atop the world's largest offshore gas field, the Northwest Dome Field, with an estimated 150 trillion cubic feet of reserves. These reserves should provide income for that country far into the next century; but as the field is just now beginning to be exploited to a significant degree, the amount of gas which will annually be withdrawn cannot now be projected, with any degree of certainty. It is clear, however, that the deposit will last for a long time.

Similarly, these dates just given for the various countries, marking the start of declining production, based on presently known reserves and current producing rates, are to some extent theoretical, for almost certainly more discoveries will be made; but also, more importantly, rates of production will change markedly. Currently, Saudi Arabia is producing at less than half the rate it could produce and that figure will surely go up, as world oil supplies eventually tighten and the Saudis can produce more oil without depressing the price.

One aspect of the petroleum interval, largely ignored by the fortunate people who daily have the generous use of oil supplies, is that there is little chance that major segments of world population will ever enter into and enjoy this time of oil, as the developed countries know it. But the figures are convincing.

Total world oil production is now somewhat less than 60 million barrels a day. It briefly reached 65.7 million in 1979, but that is probably going to be the all-time record. The oil standard of living (and consumption of oil is a good index of the standard of living, as we view it today) achieved by the United States is approximately 17 million barrels a day, used by some 250 million people. If the Chinese alone, with a population of about 1.1 billion persons, were to achieve the USA oil standard of living on a per capita basis, they would use more than 75 million barrels of oil/day, or somewhat *more* than 15 million barrels of oil/day than is produced in the entire world.

Between China and India, at the USA rate of oil consumption, more than 125 million barrels of oil would be needed each day, an impossible figure. And these calculations have left out the oil needs of all the rest of the world — Western Europe, Eastern Europe (including the Soviet Union), Japan, and more than 150 other nations. Clearly there is no possibility that major segments of the world's population will ever enjoy an oil-based economy. Human labor, and elementary energy supplies, such as biomass (wood, crop wastes, animal manure), will remain, for many years, if not for centuries to come, the basic energy sources for many regions.

If oil, however, were devoted chiefly to industrial use by China and India, the volumes of oil involved might be reduced to one-half or even one-third of the amount required to achieve an over-all USA standard of living. But even at that reduced figure, the amount needed would be in the vicinity of 40 million barrels of oil a day, still leaving very little for the rest of the world. Indeed, what would be left from world oil production would be little more than what the United States alone uses today.

Again, it is clear that in terms of oil, there are large populations which cannot become industrialized as is the western world today. In the case of China, however, there are huge deposits of coal which can be further developed, and indeed are used extensively today (with an attendant severe air pollution problem). In any event, China's current oil production of some 2 1/2 million barrels of oil a day, even if doubled, or tripled, will not lead the way to an oil-economy for them. The Chinese are said by some observers to be "essentially self-sufficient" in oil. But when you consider less than three million barrels of oil per day being divided by more than one billion Chinese, you realize they are self-sufficient in oil only in the way they use it today, not as they might use it to replace hand labor, if they had the oil available to so do.

Oil provides a great many amenities for those who have the oil. In the case of the United States, it is a weekend at the beach, a drive to the ski slopes, a trip with an outboard motor, or perhaps even a yacht over interesting waters. These things the average Chinese will never enjoy. But, how long will citizens of the United States continue to have them?

Finally, when one looks at the figures for oil reserves, and notes the wide differences in oil supplies, region by region and country by country, the question arises as to just when and how the Petroleum Interval will end. For an individual country, the end of its oil reserves may come soon or it may be a hundred years from now. But we have a global economy. Can one country enjoy abundant oil for decades beyond what other countries, previously used to ample oil supplies and with economies built on that base, can have?

The uneven geographic distribution of world oil resources, which geology has given us, presents a great challenge in what to do about this matter. Can it be solved peacefully or will there be an attempt to solve it by military action? A revision of lifestyles and the implementation of strong energy policies quite probably will eventually be in order for the western, oil-short, industrial nations. Given the time-lag involved in planning and implementing these major changes, the time to do this is immediately at hand. But human nature being what it is, the problem may be faced only on an eventual crisis basis. We would hope that even under those circumstances that actions will be reasonable and not violent. The behavior of the American public, in the long gasoline lines of the 1970s, however, is not reassuring, as even murder was committed.

Most likely, however, the end of the Petroleum Interval will be gradual, wherein no crisis point is reached, just slow change. For example, the complex of hydrocarbons which constitutes oil, has long been thought by some to be too valuable simply to burn for space heating, as is now done in home oil burners. Biochemist and former Harvard University President, James Bryant Conant, believed that by 1980 there would be a law against such a use of oil, but his guess on time was not quite right.

However, it probably will not be a matter of *if*, but *when* such legislation occurs. Space heating can be accomplished by electricity directly, or in some regions, by using electric heat pumps. Some of this power can be generated by hydro sources and much more by nuclear plants. One cannot make petrochemicals — medicines, plastics, paints, synthetic rubber, insecticides, and the myriad other very useful products which come from oil — from running water or from uranium. Thus, ultimately as oil becomes scarcer, it will gradually be used more and more for its highest value end product, which will probably be petrochemicals. So the price will inch up and little by little the lower value end uses of oil will have to use other materials. The last large general use of oil will probably be in motor vehicles, which will, at the end, be much more efficient than are those which presently exist. For that day, a new generation of cars is being planned; but then these, too, ultimately will have to be replaced by vehicles using other forms of energy than oil.

It may be noted that the use of oil today to produce plastics may actually be an economical use, insofar as the plastics are used in vehicles to reduce weight. The reduction in weight, over the years of use in a vehicle, may save more oil than was used in making the plastic. More and more plastics in automobiles of the future is almost a sure thing.

Given the increased inter-dependence of nations, and in the case of the Middle East oil producing nations, who have invested substantial amounts of their money in the western industrialized world, there may not be any more sudden cut-offs of oil supplies. And thus there will be no crisis but a gradual transition to a less and less oil-dependent world economy.

The great hope and widely-held public view is that alternative energy sources will be found to gradually fill the void left by diminishing oil supplies. In the next chapter we look realistically at these possibilities.

Chapter 11

Alternative Energy Sources: Non-renewable

Energy, it is worth stating again, is the key which unlocks all other natural resources. Nations which can continue to command substantial energy resources in the future, will be best able to survive and compete. Petroleum (oil and gas) is the most important energy source today, but this source is finite. In the decades to come, there will be large differences from today in what is termed the "energy mix"; that is, the amount the various energy sources contribute to the total energy supply. This mix is continually changing.

Until 1880, wood was the principal fuel used in the United States. From 1880 to about 1945, coal was the largest single energy source. Since that time, petroleum (oil and gas) has been the most important energy source and now constitutes about 65 percent of United States' energy. Nuclear energy has had low acceptance in the USA; and recently, development of nuclear power has slowed even more. Solar energy in certain areas — and for limited uses — is gradually being developed.

Chevron Corporation has documented the energy mix in the United States as of 1986, and made an estimate of the mix in the year 2000. (Copyright 1987, by Chevron Corporation).

U.S. Energy Consumption
% Share

	1986	2000
Oil	44	41
Gas	21	20
Coal	22	24
Hydro/Misc.	4	5
Nuclear	5	7
Solar/biomass	4	3
Total	100	100

The lesson from all of this is that energy sources wax and wane. The energy mix in the past was constantly changing in the United States and in the rest of the world; and it will continue to change. Such being the case, the question then is what alternative energy sources exist, how good they are, and who has them around the world?

As petroleum is now the leading energy source, what we regard as "alternative" energies are all other kinds of energy. This is the current conventional concept of "alternative" energies in the industrial world. Accordingly, we shall review the energy spectrum from *wood* to *fusion*. As alternative energy sources will steadily gain in importance, as there are a variety of possibilities, and as there is both a large amount of information, as well as some popular misconceptions about these sources, this section of the book is correspondingly rather lengthy. We justify this, however, on the basis that the subject is of personal importance to nearly every one now, and will be even more so in the not too distant future. It is important to know the realities of what energy prospects lie ahead. Some illusions exist in this regard; and we hope here to dispel such, and present the facts.

One might initially ask if the list is really complete, or is there some significant energy source which we do not yet see or see only dimly? It is a fair question, inasmuch as before the advent of the atomic bomb, the potential energy from the atom was only glimpsed by a few. Before Col. Drake drilled his well, and the Industrial Revolution made use of this liquid energy, the potential for oil was almost unknown and certainly not visualized at all, in terms of the way the energy in oil is used today. Before electricity was generated commercially, lightning in the sky was only an interesting phenomenon.

But we have come a very long way, in a scant few decades, in understanding what kinds of energy sources do exist. Though it might seem rash to state that we now know all the primary energy sources potentially available to us, the fact is that we do seem to have a pretty reliable and complete view of the energy spectrum: which goes from wood and other biomass, to fusion, the energy in the Sun. There is probably nothing beyond fusion which we could grasp and use, even if such existed; and between wood and fusion, there seem to be no energy sources which we have not at least sampled and evaluated in some way. Probably we know them all. We have already examined petroleum, as petroleum is now the major energy source for industrialized nations. Here we examine all other forms of energy alternatives, to see how useful they could become. Some of these we have known for a long time and petroleum superseded them — coal, for example. The question in that case is: Now that we have known petroleum, can we go back to coal and use it as an alternative to oil? Each energy source is distinctive; and although one can draw Btu's, or however one wishes to measure energy amounts, from all energy supplies, the question is how convenient is it to get that energy and use it in a particular application? Substituting coal, wood, or solar energy for the gasoline in your car can be done; but it presents some difficulties.

Energy sources can be grouped in various ways. A common approach is to divide them into those which are renewable and those which are non-renewable. It might also be noted that all energy sources ultimately come from the Sun, except nuclear energy (of which geothermal energy is also an aspect), and tidal energy, the latter being a combination of the gravitational influences of the Moon and the Sun. But coal, petroleum, shale oil (from oil shale), wood, wind, hydropower, ocean thermal gradients, and waves are all, in various ways, energy forms which have their source in the Sun.

Alternative Energy Sources

Non-renewable

Coal

Shale oil

Oil sands, tar sands, heavy oil

Nuclear (fission, fusion*)

Geothermal**

Renewable

Wood and other biomass

Hydropower

Solar energy

Tidal power

Geothermal**

Waves

Wind

Ocean thermal gradients

*As fusion can use either 2 hydrogen isotopes, one of which is in almost unlimited supply in the ocean, it might be regarded as renewable.

**Apparently derived from nuclear heat; renewable in direct use, not renewable in a practical sense, in electrical use.

There are variations on these energy sources, for example, fuel cells. But fuel cells are not an energy source in themselves but a different way of using energy materials, chiefly petroleum. The long word "magnetohydrodynamics" simply refers to a technology which can be applied to obtain electricity from coal, oil, or gas, rather than burning these under a boiler.

We now briefly look at these alternative energy sources, suggest how important they might be, and see who has them around the world. (Not all of these are minerals but as they all compete with one another, all energy sources, mineral and non-mineral, must be considered, in order to judge how important energy minerals might be in the future).

Coal. This is a form of biomass handed down to us from the geological past. In terms of total energy, coal around the world far exceeds the energy in the world's oil and gas. It was the major fuel of the industrial world in the recent past and it may come into its own again. The Soviet Union appears to hold the world's largest coal deposits, but not all of them are recoverable. It may be that the United States has the most economically recoverable coal. China is apparently third in amount of coal resources, with western Europe fourth. Australia, Africa, India, and South America follow, in that order; but the amounts of these latter countries are considerably less than those of the three world leaders: the USSR, the United States, and China. In China, even now coal is the principal fuel in use (with, one might add, a concurrent, large, air pollution problem).

The United States is running out of oil but coal remains in great abundance; and it has been estimated that when one adds up the energy in the coal, oil, and natural gas deposits of the United States, 90 percent of that total is in the form of coal. Also, these coal deposits are already located; there is no expensive exploration work involved, as in the case now for deeply hidden oil reservoirs. The coal deposits are a known quantity. Clearly coal is a potentially important alternative fuel to petroleum. However, you cannot conveniently put coal into the fuel tank of your car or truck. Coal can be liquefied but at a cost, and only by means of a huge, physical plant complex, if any great quantity of oil is to be produced from this source. South Africa has liquefied coal for years, as a source of about half the fuel for its cars and trucks; but the fuel is expensive and in relatively short supply.

The use of coal also has some substantial environmental problems, starting with the fact that underground coal mines are dangerous; and each year, a number of miners are killed and others have their health impaired. Most western coal, however, and considerable eastern coal is now mined by open-pit methods, and underground mines are becoming less common. But there is also the matter of air pollution and acid rain, when coal is used for power production. Research done on these problems has reduced them considerably but not entirely. Thus, coal has some environmental problems, if simply used by burning it, as is done now. If it were used as a substantial substitute for oil, by liquefying it, the cost of putting in place the physical plants which would be needed to supply the United States with oil, as we use it now, would be very large. And to mine the coal which would have to go into these plants would involve the largest mining operation ever seen in the world. Coal can be a substitute, in part, for oil but probably cannot completely replace oil, as we use it today. However, if coal is the coming fuel, then the Soviet Union, the United States, and China are in the driver's seat. Saudi Arabia and other Persian Gulf nations are not even faintly in the running.

Shale oil. (Oil shale) In this resource, the United States has the largest deposits in the world, with an estimated two trillion barrels of oil-equivalent hydrocarbons locked in rocks in Wyoming, Colorado, and Utah — the richest of which are in western Colorado. There have been great expectations for shale oil, particularly during the oil crises of the 1970's. Numerous fitful attempts have been made to develop a viable shale oil industry in the United States; but there are several problems: First, there is no oil in oil shale. The organic material in oil shale is called kerogen. And the material is not shale but what geologists term an organic marlstone. But as one promoter put it, "New York bankers won't invest a dime in 'organic marlstone,' but 'oil shale' is another matter." So "oil shale" is a promotional term.

To obtain what eventually (by various means) can be converted to substances something like conventional oil, the rock must first be mine (blasted out, to begin with), and then loaded on trucks or by other means hauled to a plant where it must be heated to a temperature of about 900 degrees Fahrenheit. From this, a tarry mass emerges, to which hydrogen must be added, in order to make it flow readily. Currently, the chief source of hydrogen is natural gas, which unfortunately, brings us right back to petroleum that we are trying to replace. Also, the rock, when heated, tends to pop like popcorn, so the resulting volume, even after the organic material (kerogen) is removed, is larger than the volume of rock initially mined. This makes for a large waste disposal problem. The waste material has to be hauled away to somewhere. It has been suggested that the ideal situation would be to have a mountain of oil shale near something equivalent to the Grand Canyon, where the oil shale could be brought down the mountain, largely by gravity, run through the processing plant, and then the waste material dumped into the adjacent canyon.

Developing oil shale deposits involves huge materials handling and disposal problems. Also, when the energy costs of mining, transporting, refining, and waste disposal are considered, the net amount of energy recovered from oil shale is relatively small. It does not begin to compare with the net energy reward now obtained through conventional oil well drilling and production operations. Some studies have even suggested that the final figure for shale may be negative. At best it is not large, and surface mining for oil shale may disturb up to five times as much land as would be caused by coal mining, for the same net amount of energy recovered.

Another problem in the case of western Colorado oil shale deposits is that processing and the auxiliary support facilities use water, and the richest oil shale deposits are located in the headwaters area of the Colorado River. This river now barely (if at all) makes it to the Gulf of Lower California. Present demands are more than the river can meet. Water supply would be a serious problem to any large development of oil shale.

All of these factors added to some difficult technological problems of just how to most efficiently process the shale (organic marlstone), have combined to delay development of these deposits. So far, very little oil, except on a pilot plant scale, has been produced. Chevron, Unocal, Exxon, and Occidental Petroleum all have made major efforts to develop a viable, economic, commercial operation; but none has been successful.

Unocal has been the most persistent pursuer of shale oil production and has built at least two plants in this effort. In their 1987 annual report, Unocal stated, regarding their current project: "The ultimate goal is to achieve steady production at design capacity — about 10,000 barrels a day." Scaling this up to where it would be a significant amount in terms of United States' oil consumption would be a tremendous task and, at best, many years in the future. It might be added that in 1988 Unocal considered shutting down its pilot shale oil plant, built at a cost of $650 million and located just north of Parachute, Colorado. It was operating with the aid of $400 million in Federal guaranteed subsidies, paying Unocal the difference between the market price of oil (then about $15 a barrel) and the adjusted price support of about $45 a barrel.

When oil was $5 a barrel, then $10 was the magic figure at which shale oil would be competitive, so said the industry. When oil reached $30 a barrel, the figure for shale oil was $40. Oil shale development has been "just around the corner" for the past 50 years or more, and may continue to be in that position for some time to come.

In view of the water supply problem, the absolutely enormous scale of the mining which would be needed, the relatively low (at best) net energy return, and the huge waste disposal problem, it is evident that oil shale is unlikely to yield any very significant amount of oil, as compared with the huge amounts of conventional oil now being used, in the near or even foreseeable future. Shale oil can, at the most, supply only a small portion of what are now the world's oil demands. Also, shale oil, by its composition, is better adapted for use as a raw material for petrochemical plants, than for the production of gasoline. As a petrochemical feedstock, which perhaps could justify a higher price than would a substitute for gasoline, shale oil may play a modest role in the (reasonably near) future economy.

Oil shale exists in a number of other countries, notably Brazil, Scotland, and Estonia. China also has some deposits. A modest shale oil production operation was conducted in Scotland, for a number of years. In Estonia, the oil shale is simply shoveled into furnaces beneath boilers of power plants and burned without processing. This results in a huge amount of ash to be disposed of but the operation seems to be economical enough to be useful.

Oil sands, tar sands, heavy oil. There are many gradations of oils, from those which are almost a gas, at room temperature, to very heavy tar deposits. Oil which can be recovered by conventional means (flowing or pumping) is the oil supply which the world now uses, for the most part. But here we discuss unconventional oil deposits as alternative energy sources, even though they are an evolutionary continuation, in effect, of conventional oil deposits. These unconventional oil sources include oil sands, tar sands, and heavy oil. We make the distinction between the first two and the third on the basis that the former have to be mined; and the latter can be produced, with considerable effort, from specially constructed wells with auxiliary equipment.

Oil sands and tar sands. These deposits are, in effect, ancient oil fields which have been uncovered by erosion or from which oil has migrated to the surface or near-surface, and has lost its lighter, more volatile, elements. The largest of these deposits is in northern Alberta, the so-called Athabasca oil sands, a few miles north of Fort McMurray, which contain an estimated two trillion barrels of oil in place. The deposit, in detail, consists of grains of sand, each of which has a thin film of water; and outside this water film, there is a coating of oil. By a hot water process, the oil is stripped away from the sand.

How much of this oil can be recovered economically, however, is a question, the answer to which really cannot be known precisely because economic circumstances change, as does technology. There is certainly a large amount of oil which can ultimately be recovered, but there are problems. One does not simply drill into these deposits and let the oil flow. The energy recovery efficiency of an oil well, drilled to a depth of 6,000 to 8,000 feet, as in Saudi Arabia, which subsequently flows 12,000 barrels a day, far exceeds that of the laborious process of mining the oil sand, transporting it to the processing plant, treating it to recover the oil, and then disposing of the waste sand. It is estimated that it takes the equivalent of two out of every three barrels of oil recovered to pay for all the energy and other costs involved in getting the oil from oil sand. Furthermore, the capital costs of the operation are high, as large plants have to be built and maintained; no such plants are required for an oil well, which either flows or can be pumped, to get the oil out and put it in a tank. To put the oil in a tank from the oil sands requires much energy and equipment. In the case of the Alberta deposits, the weather also is a difficult factor, as temperatures there may drop to 50 below zero Fahrenheit. Conducting mining operations in sub-zero temperatures and keeping all the equipment going, is difficult at best. Also, the quartz sand in these deposits is harder than steel and it inevitably gets into the machinery and causes problems.

At the present time, two plant complexes, the Syncrude plant (a consortium of companies, and including the government), and the Suncor operation (Sun Oil Company) produce a total of about 225,000 barrels of oil a day, which is sent by pipeline 270 miles south, to a refinery at Edmonton. But to put this in perspective, these 225,000 barrels a day are about 1/80th of the oil now being consumed daily in the United States. A third plant is now in the planning stages, which will increase this production figure by perhaps 100,000 barrels. But in any event, against total oil demand, either in the United States or in Canada, production from these oil sands is modest. Furthermore, it is Canada which has these deposits, not the United States — which is the world's biggest consumer of oil. There are almost no *oil* sand deposits in the United States. There are some *tar* sands in Utah and elsewhere. These are of small size, and the hydrocarbon is tar and not lighter oil; and therefore takes even more processing than does oil sand.

Heavy oil. There are very large deposits of heavy oil in the world, including some places in the United States. This is oil which has lost its lighter factions, or was initially a series of organic compounds which did not mature in the Earth as oil does, and never were very fluid. The two most notable areas of these deposits are in eastern Alberta and adjacent western Saskatchewan, and in eastern Venezuela. In the United States, deposits exist most notably in California.

In the Alberta/Saskatchewan area (particularly in the vicinity of Cold Lake, Alberta), wells are directionally drilled in groups, from specially constructed pads, to a depth of about 1200 to 1500 feet. Then hot water is injected for several weeks, and the wells are subsequently pumped for a time. Then the process is repeated several times; but a good deal of oil still remains in the ground after all this is done. Some oil is recovered but it is obviously more expensive than conventional oil. However, these oil deposits are now being developed and are able to marginally compete with conventional sources. There are at least 25 billion barrels (and perhaps several times that amount) in these deposits, but how much can ever be recovered economically is not known.

Generally in the past, synthetic fuel production from oil sands and tar deposits has not done very well without huge government subsidies; but the Canadian synthetic fuel industry is an exception. Perhaps it has succeeded, while others have failed, because it has an exceptionally good deposit of the raw materials. But in any case, synthetic fuel from these deposits now accounts for 11 percent of Canada's oil production.

The world's largest deposit of heavy oil is in Venezuela, variously estimated to be in the vicinity of two trillion barrels. Again, it cannot be recovered by conventional methods but will take considerable energy to produce a barrel. The profit/energy ratio will be low.

In the United States, particularly in California, there are substantial amounts of heavy oil; and some of these deposits are now being exploited, Unocal being one company so involved. As oil prices increase, more of this expensive oil will be recovered. Also important is the fact that in conventional oil fields, ordinarily less than half the oil in place is now being recovered, so with higher prices and better technology, and applying some of the technology now used to recover heavy oil, a good deal more may be produced than is now extrapolated into "conventional proven reserves." This will help stretch out oil supplies for a least some time into the future; but the flush production on which the United States enjoyed such a fast, fine ride to energy prosperity is gone. There is, however, still much oil available in various kinds of deposits, both here and abroad, but at a price, and with a considerable time lag in development, in order to put the needed equipment in place.

Nuclear energy: fission and fusion. Although both of these are nuclear energy, the sources are quite different. In fission, the source is uranium; or it could be thorium, both of which are moderately abundant in a number of places in the world. The United

States, Canada, South Africa, Namibia, and Australia have substantial deposits of uranium. Lesser amounts are known from Brazil, Niger, Sweden, and France. The Soviet Union does not publish its uranium resource statistics but is known to have some deposits; and they could be quite substantial, given the large and varied geological terrains of the USSR.

The amount of energy potentially available in the known uranium resources of the United States, if the breeder reactor approach is used, far exceeds the combined energy potential in all the oil, gas, coal, and shale oil. But nuclear energy currently has a poor reputation in the United States, especially after the accident at the Chernobyl nuclear plant, in the Soviet Union. Except for France, western Europe is having some reservations on the matter also. In 1980, Sweden voted to gradually phase out all their nuclear plants, with the last to be closed in the year 2010. (But what they will do after that, no one seems to know. It may be they will have to reconsider that decision, for at the present time, nuclear plants provide 50 percent of the country's electricity). However, in contrast, energy-short France has been vigorously pushing nuclear power and now generates about 70 percent of its electricity from the atom. Many other countries outside of western Europe are pursuing the nuclear power route, including Brazil and Japan, and others which have no large coal or petroleum supplies. (Even with large coal and petroleum resources, the Soviet Union has been pushing atomic power.). Uranium has a special attraction for energy-short countries, because the amount of energy which can be shipped in the form of uranium is very large per unit weight, compared with coal or even oil. Weight for weight, uranium has more recoverable energy than any other source, by far. When one short ton of uranium-235 is fissioned, it produces the heat equivalent of 22 billion kilowatt hours, which is the heat in approximately three billion tons of coal. Shipping energy in the form of uranium is a very efficient operation.

However, it must be observed that nuclear energy is transformed into electricity and therefore it can only substituted for whatever fuel is used to generate electric power. In many places, oil or natural gas have been, or are currently being, used to for electric power generation. In that circumstance, nuclear energy can replace petroleum. But unless electric cars are invented that are efficient and used in quantity, uranium will not be a major substitute, in effect, for gasoline; and it is hard to visualize atomic powered or battery-run airplanes. However, transport in the form of trains *can* use electric power; and as diesel fuel becomes more expensive, the electrification of the world's railroads is likely to be the trend. Indeed, the Soviet Union now has made extensive use of electricity for that purpose. Much of the Trans-Siberian Railroad has been electrified, thus saving diesel fuel for uses which electricity cannot replace, and for selling abroad to obtain foreign exchange. In the United States, some railroads, once electrified (as was the Milwaukee Road, in the west), went to diesel engines; but as oil supplies become more costly, electric trains will make a comeback, and the atom, converted to electricity, could provide the power.

In the case of fusion, the problem is not fuel supplies and radioactive contamination, but technology. It is generally believed that fusion involves creating and containing the heat of the Sun's core by the fusion of either of two isotopes from hydrogen, deuterium, or tritium. Deuterium exists in great quantities in ordinary water, and tritium can be produced in an atomic reactor from lithium, which is an element that occurs in some granitic rocks and in some underground brines. If the energy in the Pacific Ocean's deuterium was efficiently harnessed to generate power in fusion reactors, it could provide the energy needed to generate enough electricity to light the whole world for literally billions of years.

So far, however, the process of using fusion on a commercial scale has eluded some of our best scientists and engineers. At least one school of thought believes that the chances of ever doing this are no better than 50-50.

Note must be taken, however, of the claimed success of a so-called "cold fusion" process, reported in March, 1989 by two scientists, one a chemist, the other a physicist. Using an apparatus consisting of palladium and platinum metals, immersed in water rich in deuterium, these scientists claimed that cold fusion had been accomplished and the amount of energy emitted from the experiment was four times the amount of electrical energy which had been put into the experiment.

This announcement, needless to say, has created a great stir among physical scientists and energy experts. At the present time, the jury is still out on the validity of the experiment. Chemists claim it is an example of a new form of chemical reaction. Physicists largely support the idea that it is true fusion. Intense study of the matter is now being done at hundreds of laboratories throughout the world. One would assume that shortly the truth of the matter will be found.

If it is proven that this is true fusion, this would subsequently impact the world energy mix very considerably. However, the time it would take to "scale up" this process to commercial size would also be very considerable. A rough estimate, but perhaps on the conservative side, is that it would take at least 20 years to scale it up, build plants, and make this energy source an appreciable factor in the energy supply of the United States. To bring it to lesser developed nations, which currently use oil by which to generate electricity, would probably take even longer.

Thus, it seems that the discussion and evaluation of energy supplies can continue, based on the presently known and utilized energy sources. It might be added that even if fusion is accomplished as claimed, and can be "scaled up" to commercial size, the widespread application of electrical energy to the field of transportation — that is, to motor vehicles — is in itself a huge technological challenge, the solution to which is not immediately in sight. It would involve probably a complete revision of the fuel supply and distribution system — for longer range vehicles, it would involve battery exchange stations along highways and ultimately penetrating into the byways, as well. The transition would proceed slowly at best, because contemporary motor vehicles (hundreds of millions) now powered by gasoline or diesel fuel, would have to be phased out. For more remote regions of the world, this may not be practical for decades to come.

Oil will continue to be in demand for many years. OPEC can have little fear that its product will go out of style. But the energy potential for fusion is so large that the quest must be pursued. However, it is unlikely that fusion will be a commercial reality, on any appreciable scale, for at least several decades. The fusion route to energy will be pursued as an international, long-range goal, for no new technology on the horizon offers, theoretically at least, the energy potential of this process. Also, unlike fission, fusion does not have the risk of a runaway reaction and its by-products do not pose the disposal problems to the extent that fission does. Also, as already noted, fusion uses an isotope of hydrogen which exists in the ocean in almost unlimited quantities. So for all practical purposes, fusion may be regarded as a renewable energy source.

But to go back to fission, the nuclear energy technology we now use: If that route continues to be pursued, then a number of countries have uranium resources to obtain energy from this source, in great quantities. The decision is currently a political/environmental one. Even now, however, in the United States, 17 percent of the electric power is nuclear in origin. Such places as New England and the City of Chicago get more than half their electricity from the atom, as does the Province of Ontario, Canada. With coal the only other substantial present alternative for the generation of electricity but with the acid rain, the ozone, and the "greenhouse effect" problems of coal-burning, the whole matter of nuclear power and the production of electricity may have to be reconsidered.

Like it or not, the world, in general, is going nuclear. In 1960, the atom provided less than one percent of the world's electricity. By 1986, it provided more than 15 percent. The London-based Uranium Institute notes that in 1988 17 percent of the world's electricity came from the atom, and states that by the year 2000, the world's nuclear-derived electric generating capacity will rise from the present 261,000 megawatts to 347,000 megawatts. The International Atomic Energy Agency predicts that nuclear energy will provide at least 20 percent of the world's electricity by the year 2000. In such countries as Thailand, Turkey, Argentina, Brazil, Bulgaria, Japan, France, and Belgium, nuclear energy is in use and being pushed. As of 1988, outside of the United States nearly 600 nuclear plants, in 41 countries, were in various stages of development or in operation. Even after the Russian Chernobyl nuclear plant accident, the Soviets continued construction on 15 atomic power reactors.

As oil and gas supplies diminish, it appears there are only two sources of energy large enough, potentially, to take up the slack; these are coal and the atom (be it fission or fusion). However, due to increasing concern about the "greenhouse effect," caused by burning fossil fuels like coal (which now produces 37 percent of the world's electricity), the nuclear option is starting to be looked upon more favorably by some environmentalists. A new type of nuclear power plant, the integral fast reactor, now being developed, produces much less toxic waste than present plants and reprocesses most of its own hazardous fission products. This may be a satisfactory route to further nuclear power development, if fusion proves to be yet unattainable.

Again, in terms of uranium (for fission), the United States, Canada, South Africa, Namibia (Southwest Africa), and Australia are exceptionally well endowed with this energy resource. There are tremendous deposits in Saskatchewan that are just beginning to be developed, along with large mines already opened there and elsewhere in Canada.

Geothermal energy. This interesting energy source was first used to light a light bulb in Italy in 1905. Geothermal energy is heat from the Earth, apparently generated, in most part, by radioactivity — the decay of radioactive minerals. This natural heat occurs everywhere in the Earth but only in certain places is it concentrated enough and hot enough, close to the Earth's surface, to be commercially tapped. These areas tend to occur chiefly along the junctions of what we now know to be plate margins, as defined by the theory of plate tectonics; but there are certain so-called "hot spots" (such as the Island of Hawaii) which occur within a given plate. In general, most of the geothermal areas tend to be related to regions with active volcanoes.

Volcanoes and petroleum do not co-exist very well, so this means that, in general, areas which have many volcanoes have little or no oil, but do have considerable geothermal potential. These areas include the northern California Coast Range, the Cascade Mountains of Oregon and perhaps the southern part of Washington (Mount St. Helens, in 1980, provided a spectacular example of geothermal energy in action), the Imperial Valley of California and extending into northern Mexico, Central America, Hawaii, Iceland, Indonesia, New Zealand, the Philippines, Japan, Italy, and eastern Africa. (One estimate is that there is enough geothermal energy in the east African Rift Zone and including Ethiopia, if properly harnessed, to light all of Africa.).

Some of these places already have been developed as commercial geothermal energy sources. The world's largest geothermal plant (many plants, actually) is at The Geysers, about 70 miles north of San Francisco. Electricity was first produced here in 1960. Now some 2000 megawatts are being generated. This is about the equivalent of two good-sized nuclear plants or two Bonneville dams, and is enough to supply the electric power needs of about two million people.

Geothermal energy can be used in two ways. It can be used directly, for space heating (pipe hot water from wells and run it through radiators or heat exchangers), or it can be converted to electricity. Only the hotter geothermal reservoirs (generally above 360 degrees Fahrenheit) can be efficiently used to produce electric power; the lower-temperature waters are used directly in space heating. Direct use in space heating is the more efficient use. At the present time, 60 percent of Iceland is heated in this manner, essentially pollution free.

Klamath Falls, Oregon has long been known as a "city in hot water," because of its use of geothermal water for heating. In Klamath Falls, one can enjoy a cradle to grave geothermal life. One can be born in a geothermally heated hospital, be geothermally heated in schools (from grade schools through Oregon Institute of Technology), get married in a geothermally heated church, and live in a geothermally heated house, built from wood purchased in a lumberyard which dried the lumber, using geothermal energy. You can buy a car from a geothermally heated automobile agency and park it on your geothermally heated driveway. You can drive the car on geothermally heated pavement, and take the car to a carwash using geothermal water. You can get medical attention in a geothermally heated doctor's office, and do your clothes in a geothermally heated laundromat. If you are of good character, you can enjoy the facilities of the local geothermally heated YMCA; if not, you may wind up in a geothermally heated Klamath Falls jail. Winter snow and ice are repulsed on geothermally heated sidewalks.

For later years, there is a geothermally heated nursing home available, and, finally, one may be buried out of a geothermally heated mortuary.

Many other places also employ geothermal water for space heating. Boise, Idaho heats government buildings and a mall geothermally. Greenhouses in Idaho are geothermally heated, as are greenhouses and a mushroom-growing operation in Oregon. France and the Soviet Union make considerable use of geothermal waters for space heating.

At several other locations in the world, besides The Geysers field in northern California, already cited, geothermal energy is being converted to electrical energy, as in New Zealand, where 10 percent of that country's electricity comes from generating plants using the heat of the Earth as the power source. Italy, the pioneer in geothermal energy use, now has developed enough power to provide the equivalent of what it takes to run all the electric trains in Italy; and almost all Italian trains are now electrified. This is a great saving to Italy, which has very little oil and would otherwise have to import large amounts of diesel fuel to run its railroads. The Philippines are now using geothermally generated electricity and are planning a very large expansion of this energy source. With almost no oil, geothermal power is proving a great boon to that volcanic island country. Geothermal energy tends to be rather site specific. That is, unless it is hot enough to be used to generate electricity, which can be transported to distant areas by power lines, it has to be used fairly close to where it exists in nature. It cannot be shipped across oceans, as can coal, oil, or uranium. Thus, locally, geothermal energy can be, and is, important; but as a help to the world's energy needs, it can have only minor impact. Still, there is considerable geothermal energy that can be developed. Both Central America and the East African Rift have very large geothermal energy potentials. However, geothermal energy, at best, is a small figure in the large energy numbers needed to power the world as we know it today.

Finally, as a technical note, we have classified geothermal energy as a non-renewable resource. When geothermal energy is used for electric power generation, in the fairly short term, it is non-renewable, in the sense that ultimately the reservoirs of steam and/or hot water will be depleted to the point where they are no longer capable of sustaining electric power generation. Although that time may be delayed, to some extent, by reinjection procedures, the time to depletion is variously estimated to be from 40 to 100 years in most geothermal electric power fields. However, after being shut down, ultimately the field (over a period of many hundreds or perhaps thousands of years) will recover and could be produced again, for the heat will still be there; it is only the hydro system which gathers the heat from the fractured hot rocks and brings it to the well bore, which becomes exhausted. For this reason, in a practical sense, geothermal energy for electric power generation is not a renewable resource; but there is one technology, under testing, which may modify this conclusion. In some geothermal fields, waste water is being injected into the reservoir, to see if the reservoir level and pressure can be maintained, without adversely reducing the temperature. Results are not yet conclusive; but if this is at least partially successful, then geothermal energy for electric power will be at least partially renewable in practical terms.

However, when used in space heating, application of the hot water may possibly be controlled, so as to be kept in balance with the recharge of the hydro system, which brings the heat from the fractured hot rocks to the well bore. In this case, geothermal energy is a renewable resource. Thus, depending on its end use, geothermal energy can be thought of either as renewable or non-renewable. In the case of space heating, however, even here the reservoir can become a depleting situation, if it is overused. Such seems to be the case in several of the district heating systems in southern Idaho, where a study of this problem is underway. However, use of geothermal waters for space heating is the most efficient use of such energy and should be pursued. Even where there is no especially warm water to use, the general heat flow of the Earth can be drawn upon by means of groundwater heat pumps, which are efficient in most temperate areas, and better in some places than the usual air to air heat pumps.

Chapter 12

Alternative Energy Resources: Renewable

We have been considering non-renewable energy resources. Obviously, for the very long term, we must depend on renewable energy sources. As we have already noted, for the most part these are not regarded as minerals; but as they compete with energy minerals, the economics, the availability, and the dependability of these renewable energy sources must be considered along with their mineral rivals. These we now review.

Wood and other biomass. Until about 1880, wood was the chief fuel used in the United States and still is the principal fuel in many parts of the world. But the use of wood is deforesting many areas, causing huge and fatal landslides, devastating floods (as in Bangladesh, in 1988), and widespread erosion and loss of valuable topsoil. In Haiti, where the country was at one time nearly all forested, now only two percent of the land is wooded, and the demand for charcoal, which is the chief fuel for the majority of the population, exceeds the reforestation rate. Topsoil is being lost rapidly. As one flies into the Haitian capital of Port au Prince, the brown ring of muddy water, fringing that island country, testifies to the disaster that a lack of something else (besides wood, as an energy source) has become for that impoverished land.

In areas such as the countryside around Bogota, Colombia, the highlands of Peru, and throughout much of India, Pakistan, Bangladesh, Nepal, and in several other regions, there is a very serious, crisis in the supply of firewood. It is doubtful if the use of wood for fuel can be greatly expanded from its present volume worldwide; and in fact, in many regions the forest should be allowed to grow back, to prevent further erosion and floods, which already are a severe problem in some deforested areas.

Other biomass besides wood is also used, such as wastes from crops (for example, sugar cane stalks, used to fire boilers at sugar refineries), garbage, and animal excrement. Locally, these sources can be important to the population; but they are not high-grade energy sources, and probably cannot substitute, in any great amount, for other energy supplies now in use. The conversion of wood to alcohol has been done; but the economics are marginal (that is, the net energy thus obtained is small). The amount of fuel which could be derived from this source is limited; and again there is the problem of deforestation, if wood were used in an extensive way for conversion to alcohol fuel.

In regard to alcohol, special mention should be made of ethanol, which is alcohol made from grain — commonly corn. At, present, some oil companies blend ethanol into their gasoline products, for various reasons (the mix commonly called gasohol), but chiefly because it raises the octane rating. The corn-producing states such as Iowa and Nebraska are strong advocates of this source of fuel for the nation, as are politicians who are running for office in those areas. Even presidential candidates embrace gasohol while campaigning in the corn states, being either ignorant of, or simply ignoring, the fact that the economics are not attractive, or even positive, in terms of net energy recovered. One might parenthetically note at this point that the United States can no longer afford politics as usual, where energy and minerals are concerned. The date of that transition was 1970 when the United States could no longer supply its own oil needs. The surplus "cushion" allowing for indecision on implementing a strong national energy program is gone and the realities should be faced. A long-term, comprehensive energy policy for the United States should be drawn up and vigorously pursued; but in no national campaign so far has such a fundamental thing been seriously discussed — a rather sad commentary on the American political scene.

To return to ethanol, at present what makes it economically possible to put ethanol in gasoline is chiefly the subsidy or tax rebate that oil companies and independent blenders of gasoline products get when they mix ethanol with ordinary unleaded gasoline. This is a political matter again, for without this federal subsidy, given corn even at the low price of $2 a bushel, the price of oil would have to be at least $40 a barrel before ethanol would be competitive. For comparison, oil is currently priced between about $18 and $20 a barrel.

We may see $40-a-barrel oil before the end of the century, but there is another basic problem with ethanol from corn (the grain source most widely favored for ethanol production in the United States). If the energy cost of plowing, planting, fertilizing, and harvesting the corn, together with transporting the corn to the processing plant and the energy involved in processing the corn to alcohol are all added in, the net energy result is negative. That is, less energy is recovered than it took to produce the ethanol. I am advised by Iowa State University engineering agricultural economists (and they should know) that only if the corn is delivered to the processing plant, with all previous energy costs paid, can ethanol be produced from corn with a net positive energy recovery. It should also be noted that ammonia fertilizer, widely used on cornfields, is now made from natural gas and therefore becomes a rather direct energy cost in growing this grain.

A longer range cost, difficult to assess in precise terms, is that growing corn is hard on the land and there is a potential loss of soil fertility from years of corn cropping and from the resulting, inevitable erosion of the land.

Also, in terms of the total resources which go into the production of corn, it is estimated to take about 3600 gallons of water to produce a bushel of corn, and some of this corn is produced by irrigation, which involves pumps that use energy. Also, in many areas, the aquifers are being over-pumped, as is the Ogallala Formation, which underlies parts of eight midwest to western states. The total costs of producing ethanol are high. It is a vastly different set of economics from drilling an oil well and then simply opening the valve and letting the oil flow out.

Finally, in the case of all biomass, the total volumes of raw materials needed to produce enough alcohol to make a substantial impact (in substituting it for oil) are really not there. Locally, alcohol thus produced has helped; but it is only marginally economic (as in the case of Brazil, which has been producing alcohol from waste sugar cane stalks). But in larger terms, alcohol at best could supply only a small fraction of the world's mobile energy supplies (gasoline, diesel). The costs would be high, some of them direct and immediate, and others, as we have noted, more subtle and longer term, such as erosion and loss of soil fertility. And, in fact, these long-term costs may ultimately be higher and more serious than the more immediate costs. Also important is the fact that production of ethanol, on any significant scale, would involve taking out large areas of agricultural land currently devoted to food production. In many countries such land could not be spared to produce ethanol. It must be used to feed the population.

It should be noted that alcohol as a mobile fuel has some disadvantages. A recent test of methanol, made either from coal, wood, or natural gas and a close alcohol cousin to ethanol, was run by United Parcel Service, which has a very large fleet of delivery vans. The experiment was done in southern California, and a blend of methanol and unleaded gasoline was used in 50 vans. The results were higher fuel costs, poor vehicle performance, more maintenance, frequent breakdowns, and lower mileage.

In 1988, however, Japan and China, in a rare show of cooperation between these historically long-time enemies, agreed to work jointly in developing a methanol from coal technology. The logic of this is that both countries see that they will eventually have to go to some other fuel than gasoline; China has the coal and Japan has the technology. In this instance, a potential energy mineral shortage has resulted in cooperation, rather than division, between nations.

Hydropower. The Sun initially heats the ocean water to evaporate it, and then the wind, caused by the uneven heating of the Earth's surface by the Sun, carries the water vapor to the land where, as the atmosphere rises, clouds are formed and moisture comes down chiefly as rain or snow. This fills streams and is stored by dams and is ultimately run through turbines which power generators to produce electric power. Thus, hydropower is derived from the Sun as are virtually all other energy sources. This is an important and economic source of energy, widely used.

When the technology for electrical power generation was first developed, no rivers of any size had dams across them (only some small streams were dammed, to run local grain mills). The big power dams which now exist are relatively new. Even Bonneville Dam, the first impoundment to provide power to be marketed over the Bonneville Power System, did not produce power until 1937. The initial phase of Grand Coulee Dam was not completed until 1950. And the final phases of construction on some of the dams (and even on Bonneville) were not completed until just the past few years on the Pacific Northwest Columbia River power and irrigation complex. Thus, the big dams are a relatively new arrival on the scene and their ultimate impact on a region has yet to be determined. However, the completion of the huge Aswan dam on the upper Nile, in Egypt, has already demonstrated some of the adverse effects which such projects might create.

This dam does, indeed, control the annual Nile floods, which it was designed in part to do. But these flood waters were very useful, for they left behind layers of new, fertile soil each year, in the very important agricultural area of the lower Nile valley. Furthermore, the land which annually used to receive the fertile Nile sediment now has to be fertilized artificially; and the peasant farmers generally do not have money to buy such fertilizer. The annual floods also washed away salts from the soil, which would otherwise injure plant life; and they swept away the snails that carry the Bilharzia larvae, which cause a very debilitating or fatal intestinal infection, schistosomiasis, in humans.

Due to the fact that the silt of the Nile is now trapped behind the dam, the Nile delta in the Mediterranean, which formerly was building outward, is now being eroded. This is an exceedingly productive agricultural area, and about one-third of the delta edge is now being cut back at the rate of as much as two meters a year. The Nile floods also brought nutrients into the eastern Mediterranean; but now, without these the plankton populations, the start of the food chain has been reduced, with the result that crustaceans, mackerel, and sardine populations have also shrunk, and some 30,000 Egyptian fishermen have lost their livelihood.

The dam does provide irrigation by canal to some 2,800 square kilometers of formerly desert land; but the canals carry the Bilharzia larvae, and as a result, schistosomiasis infection has greatly increased. Analyzing the various factors involved, some observers believe the dam and reservoir are, in total, an ecological disaster (Griggs and Gilchrist, 1977).

Thus dams which are commonly thought of as environmentally clean sources of power, actually may have many negative consequences. Dams flood fertile lowlands (as in Kansas, where billboards chronicled the protest "Stop this big dam foolishness"). In more northerly regions and in mountainous areas, river valleys are important winter rangelands for wildlife, particularly big game animals. When these valleys are filled with water, they no longer are winter range.

Looking longer term, ultimately all reservoirs will silt up. Hoover Dam, at the lower end of Lake Mead, will eventually become a great concrete waterfall unless the silt behind the dam is flushed downstream which, however, simply moves the problem to another site. The silting up of Lake Mead, behind Hoover Dam, is already clearly visible for several miles along the upper end of the lake, where it is backed up into the Grand Canyon, and it is not a pretty sight. However, these problems will not be acute for many years to come; and we do not now give any great thought to the matter. Let the future generations solve the problem! We can enjoy the benefits of these dams now. Hydropower does have such large advantages that we simply ignore the ultimate problems it may leave. And then there is always the possibility that a happy answer will be found in the interim, although, so far, there has been no solution to the smaller, mud-filled reservoirs, which already exist in some areas.

As to who has the hydropower, the resource is widespread. The United States has a modest amount, but nearly all of the large sites, and many smaller dam sites, have now been developed. Like its oil, the United States has already used many of its superior dam site resources. Canada, however, still has a large number of excellent, unused sites. Some of which, however, would involve destruction of important wildlife habitat, if developed for power. South America has a considerable amount of undeveloped hydropower; but much of it is along the eastern flank of the Andes Mountains, in rather remote areas. The Soviet Union, which has done a great deal with large, low-head dams, has enough undeveloped hydropower to serve the needs of an estimated 100 million more people. Africa has the largest resources of undeveloped hydropower. Japan, Australia, and Europe have small amounts; Norway having relatively more than others. In the case of Switzerland, which has within its borders the higher gradient headwaters of some western European streams, electric power is one of its more important exports but the amount is small. China has a fair amount of undeveloped hydro resources, but not as much as one might expect, from the size of the country. Also, large developments would effectively put out of production what are now some very important and fertile agricultural lowlands.

Thus Canada, along with Brazil, Argentina, the Soviet Union and certain countries in Africa are the chief holders of major hydropower sites. These will be important to those countries in the future.

Hydropower is classed as a renewable energy source, but the inevitable silting up of the reservoirs, although it may take hundreds of years, adds a thoughtful note to the concept of hydropower as continually renewable. For the present and past few generations of people who have had sites on which to build dams, the resource has been renewable, as far as they are concerned. Each year brings the annual rain and snow to refill the reservoirs, but the water also brings in sediment, which stays. The fact that, due to silting, the reservoir capacity is slightly less each year can be ignored for the time being. But for the generations to come, with no new hydro sites to occupy, these existing sites — gradually filling with sediment — may not represent a renewable resource, but a large, muddy problem.

Solar energy. By this we mean collecting solar energy (by some method) directly from the Sun and using it. Solar energy is so obvious and so universal that it has very wide appeal to many people, as an abundant and free energy source. It also is pollution free — there are no radioactive end products, no acid rain, and no waste disposal problems. It looks so good to many environmentalists, in particular, that it has been said (with some degree of truth) that solar energy is no longer a science, but a religion. Some people believe in it as the great ultimate solution to our energy supply problems; others are somewhat skeptical.

The raw numbers for solar energy tend to be impressive at first glance. Some 15,000 square miles of the United States, properly located, if covered by solar cells, would provide all of the electricity the United States currently needs. But the practicalities are another matter. There is the problem of storing the electricity, and the matter of servicing the energy gathering equipment. Perhaps these difficulties can be overcome in time; but the costs of producing electricity from the Sun are presently quite high, although some progress is being made to make solar cells more efficient.

In remote areas, and in satellites, where cost is not a consideration, solar energy can be quite useful. And in small, local uses it can be economical — e.g., in powering relay stations for message transmissions, and in the form of solar water heaters for domestic use. Israel has many solar water heaters now in use. Heating houses entirely by solar energy has been more difficult; but proper construction of a house and orienting it to take best advantage of the local solar situation, can be very helpful in obtaining at least part of the house heat from the Sun.

In an attempt to develop solar energy for commercial applications, against other energy sources, particularly in developed areas such as the United States, a number of experimental projects have been undertaken. At Solar 1, Southern California Edison's experiment near Daggett, California, in the Mohave Desert, several thousand mirrors focus the Sun's rays on a large, liquid-filled tank, raising the temperature to above the boiling point and producing electric power by means of a steam turbine and generator. There is also a superheated, fluid storage arrangement, which does allow some power to be generated for a few hours after the Sun goes down. Cost of the electric power thus produced, however, is still higher than that generated by conventional means.

ARCO, the oil company, at one time was a major producer of photovoltaic (solar) cells but apparently decided that payoff was too far in the future and in 1989 sold their operations to Siemens. Photovoltaic cells produce electricity directly from sunlight. Experimental banks of cells have been in operation in the Mohave Desert, near Hesperia, California for some time. A variety of efficiencies have been achieved with experimental solar cells. Most range from 10 to 20 percent; but in 1988, scientists at Sandia National Laboratories built a cell which is 31 percent efficient. In 1989, Boeing achieved 37 percent.

Solar energy will no doubt increase its share of the energy mix in the future. However, this is likely to come slowly. There is a trade-off in regard to solar cell cost and efficiency.

The most efficient cells cost considerably more than the less efficient cells. Thus, it may be just as cost-effective to use low-cost, low-efficiency cells, than to use the high-cost, high-efficiency cells. At the present time, solar cell systems produce electricity at a cost of from about 30 to 40 cents a kilowatt hour. However, it is predicted that the new generation of solar cells, called the "thin film" cells, will bring down the cost to less than 15 cents a kilowatt hour. This would be within striking range of a commercial situation.

As to who has solar energy around the world, here it is a case of the rich getting richer, as the Arabs clearly have been blessed in this regard, also. The Persian Gulf nations have abundant solar energy, as does Egypt and the other countries along the north coast of Africa, all the way to Morocco. One might also cite the now (relatively) economically useless Sahara and central Australian deserts as great solar energy resources.

But, there are many other areas of the world with a high solar energy input. Most everywhere on the Earth there are times when at least a modest amount of solar energy is available. How much the presence of abundant solar energy, or lack thereof, might affect the future of a nation can hardly be visualized at this time. Much will depend on further developments in solar energy technology. However, as solar energy most surely will become increasingly important, the countries which have it in quantity and quality will no doubt benefit. Will an electric power grid from solar collectors on the Arabian Peninsula someday bring heat and light to Europe, even as Saudi oil does today? It might also be noted that a major shift to a solar energy economy, from our present oil-based economy, would almost certainly involve some large changes in life-styles and a considerable reorganization of society.

Tidal power. Watching the surge of tide in bays and estuaries, one cannot but be impressed with the amount of energy involved in moving all that water twice a day. Harnessing the tides, however, has been a rather elusive matter. Tidal power sites are quite limited in number. First of all, the location must have a tidal range in the vicinity of 25 feet or more, if it is to function satisfactorily, for there must be a fairly strong flow of water through the turbines. These sites, for the most part, are in high north and south latitudes, generally at considerable distances from where the power would be used. An example is Frobisher Bay in southern Baffin Island, west of Greenland, where I witnessed a 50-foot tide. It was impressive, but a long way to where power, in quantity, would be needed. The second requirement for a viable tidal power site is a favorable configuration of the land. There must be an estuary into which the tide will funnel and where it can be trapped by a dam, wherein turbines are installed which can work both ways, on an incoming tide and an outgoing tide.

There are apparently less than a dozen such sites in the world; and, again, most of them are in fairly remote locations. At present, there are currently three places where tidal power has been developed. One is in the northwestern portion of the Soviet Union, at a site with the nearly unpronounceable name of Kislayagobuga, near the city of Murmansk, on the an inlet of the Barents Sea. One is on the Rance River in France. The third is on Hog's Island, near the mouth of the Annapolis River in Nova Scotia, Canada. This plant is somewhat different from the other two, in that the tidal range is only from about 14.5 feet (neap tides) to 28.5 feet (spring tides), with the average being about 21 feet. Power is generated only during the out-going tide. This station is unique because of the relatively low head (drop of water) by which it operates. It is claimed to be operative in a range of 4.6 feet to about 22 feet. To do this it uses a new type of low-head turbine called Straflo, developed by Escher Wyss and installed first in 14 low-head, submerged stations in South Germany and Austria.

The project is still under economic evaluation. If the economics seem reasonable, given what the cost of power may be in the future, this sort of plant might open up substantially more areas for tidal power development. However, at present the cost of power at all the plants is fairly high, although the French Rance River operation appears to be successful and competitive with other forms of energy there. Tidal power economics are definitely the best where the tidal ranges are large; and this factor, combined with the need for a special configuration of the land, does limit the number of potential sites.

Furthermore, as tidal power is not a continual power source over a 24-hour period, but can be produced at only certain stages of the tide (and these tidal stages change in time from day to day), there is a problem of integrating this sort of a power schedule into a large, regular power grid. Tidal power may, in selected areas, be useful. At best, however, it is unlikely that tidal power will be a very significant element in the total world energy picture.

Waves. Along with the tides, the strength of waves is impressive; but, like the tides, harnessing this energy is difficult. At times, waves can be 60 feet or more in height; and yet, at the same location, for weeks the ocean may be essentially calm. All sorts of mechanisms have been devised in an effort to usefully capture wave energy. The ingenious Japanese, surrounded by waves and very short of energy, have been interested in this matter; but so far, I am advised, the best they have been able to do is light a buoy at the mouth of Tokyo Bay (part of the time) by a wave-energy, electric-conversion device. Wave energy is not likely to be a factor in the world's future energy picture.

Wind. The Dutch did not invent the windmill, but were relative late comers in their use. For a time, they did make an asset of the miserably cold winds which blow in from the North Sea, by constructing windmills; but very few, if any, are in use at present. The Persians, many centuries B.C., were the first to use windmills and employed them in pumping water for their arid land, now known as Iran. The windmill has kept the western United States rancher in business, by pumping water for his cattle, scattered across thousands of acres of dry range land. With the aid of a large number of storage batteries, windmills in the past also provided the rancher (and other remote people) with some stored electric power.

More recently, windmills have been syndicated, and the concept of "wind farms" has given rise to limited partnerships, which, in turn, built windmills and sold the power to power companies; but these projects were subsidized by the federal government in the form of tax write-offs. Without them, windpower would not have been economical, at least at this time.

These windfarms also are not universally admired, especially by the citizens who may live near them. Ask the people on the western edge of Palm Springs, California about this situation. Windmills are noisy — very noisy. And they take a great deal of maintenance and can easily be damaged by the vagaries of the wind, which at 50 or more miles per hour can change direction 90 degrees, in less than five seconds. This is hard on equipment.

Wind energy can be locally important, but at a considerable price. Also, like the Sun, which doesn't shine all the time, the wind, which is caused by the heating of the Earth's surface by the Sun, doesn't blow all the time. It is undependable and is not likely to be a significant energy source in the foreseeable future. Locally, however, in more consistently windy areas, such as in Denmark, which is very short of indigenous energy supplies, windpower is being used. Britain's Department of Energy and Central Electricity Generating Board, in 1989 launched a wind park program, estimating that wind could contribute up to 20 percent of Britain's electricity. However, it was calculated that with present technology it could take one hundred square miles of windmills to produce only one percent of the country's electricity.

In terms of any large-scale commercial use of windpower, however, so far, in the United States, at least, windfarms have produced mostly experience and some fairly costly electricity which, without tax credits, would not have been produced at all.

Ocean thermal gradients. This is also called Ocean Thermal Electric Conversion or OTEC. The principle is this: When there is a temperature difference between two masses, there is a potential for power generation. In the tropics, for example, the surface water of the ocean is warm, but the depths are cool. The warm water can be used to evaporate a low boiling point fluid, such as ammonia; and the cool water can be used to condense it. However, as the temperature differential in the ocean is not great, the efficiency of this operation is low; and huge volumes of water have to be circulated to generate any significant amount of power. This has been done experimentally on the Kona Coast of Hawaii, and on a small island off the coast of Brazil. It has not been done on a power plant anchored in the open ocean, as some have suggested. The problems of building such a plant are very formidable, and the economics at the present time do not look attractive. Maintenance costs against storms and corrosion are likely to be high. It does not appear that this source of power will be developed to any extent in the foreseeable future.

For the sake of completeness, hydrogen, sometimes suggested as a great new fuel for the future in the context of a visualized "hydrogen economy," should be mentioned. However, hydrogen is not a primary source of energy, in that it has to be produced by means of other energy sources (commonly by the electrolysis of water, or derived by processing natural gas). Therefore, whereas hydrogen burns — with the end result being non-polluting water — and is environmentally attractive, it has to be produced from an energy source which we have already considered. Again, it is not a primary energy source; the most common concept in large scale production of hydrogen would be from water, by use of nuclear-generated electricity.

This has been a fairly exhaustive survey of energy sources; but energy is so vital to everyone that a realistic view of the prospects seemed in order here. One of the conclusions which comes out of this is that there is no single, adequate energy substitute on the horizon, for the conventionally produced quantities of oil and gas which we use in the world today. Also, replacing petroleum by other energy sources, at least to any large degree, will involve very large amounts of capital, and will probably cause considerable changes in lifestyles. Furthermore, once oil is no longer available in large and relatively cheap quantities, which can be conveniently shipped around the world as needed, such countries as Japan will be severely impacted, since they have no significant indigenous energy sources on their islands. At the present time, Japan is dependent on an oceanic highway of tankers, about one oil tanker every 100 miles, strung out "along the road" from the Persian Gulf to the Japanese islands; a continuous procession that goes on every day of the year, and year after year. Eventually, of course, it will stop. Then what?

It is likely that atomic power is the route Japan will have to pursue; and, indeed, they are already making extensive use of the atom. Japan is in a difficult situation in regard to energy and it may be the first country to illustrate, albeit reluctantly, what happens to an industrial nation which has no significant domestic energy base. In some fashion, after oil, they must continue to import energy.

However, because Japan is already well along on the nuclear power road, there is one school of thought which suggests that beyond the oil era, Japan will do even better than they are doing now, relative to other nations in the matter of energy. By being forced early to "go nuclear" (which appears, along with coal, the only other source of energy on the horizon large enough to replace petroleum), they will have pioneered the use of such power nation-wide.

An analogy already exists in that Japan, because of having no oil, developed smaller and more fuel efficient cars before the rest of the world did. As a result, Japan was ahead of other car makers (most notably those in the USA) when the price of oil went up dramatically and the time of smaller, fuel efficient cars arrived. In any case, Japan's successes and/or failures, in regard to ultimately replacing petroleum as the nation's chief fuel, will be followed with keen interest by the rest of the world. But it should be noted that to go nuclear, Japan still has to import the basic energy source — uranium. Japan does not have it.

The energy mix has changed markedly the past two hundred years, for the world as a whole. Some areas, however, still use energy sources (biomass largely) which they have used for hundreds, if not thousands, of years. So, in kinds of energy sources being used, parts of the world are centuries apart from other regions. This disparity will probably continue.

The survival of countries in the future, with regard to energy (and this will have a lot to do with how well they survive, in general), must involve an increasing amount of technology, combined with such energy resources as they can obtain.

It seems clear also that the future will involve a less concentrated and more diffuse "mix" of energy sources than at present. Petroleum now carries a very large load in the industrial world; but as this energy form is gradually depleted, that portion of the energy demand which petroleum now supplies probably will be gradually taken over by *several*, rather than one, other energy source. This could actually result in healthier economies, as no one supply would be so critical, as is petroleum today. Perhaps then nations will not be so vulnerable to cut-offs of energy supplies as has been the case in the recent past.

If one takes the very long view of the energy situation, *renewable energy resources* presumably must be the ultimate energy supplies. But in the interim, which probably means several centuries to come, other energy minerals will still be important as alternatives to the major role which petroleum now plays. Coal, shale oil, oil sands, tar sands, heavy oil, geothermal resources, atomic minerals (chiefly uranium and thorium), and the non-minerals — hydropower, biomass, and solar energy, will all be utilized to a greater or lesser extent, and will affect the economies of nations. The energy mix for any given nation, and for the world, will continue to be in flux, offering many challenges to both technology and world economics in the process, as does the energy mix even of this moment.

For the United States, the world's largest energy consumer, the matter of developing alternative energy sources is becoming critical, with little progress being made. Senator Mark O. Hatfield, ranking minority member of the Senate Committee on Energy and Natural Resources, stated in 1990: "Current debates over where and how to drill for oil in this country soon may be rendered irrelevant by a nation desperate to maintain its quality of life and economic productivity. Indeed, war over access to the diminishing supply of oil may be inevitable unless the United States and other countries act now to develop alternatives to their dependence on oil."

Chapter 13

Ocean Minerals and Nations

There are two categories of nations, in regard to minerals from the ocean. There are nations which have a coastline and thereby extend their territorial rights out into the sea. Earlier these rights commonly extended for three miles, which was the range of cannons in those days. Now, usually, these territorial rights have been moved out to 200 miles.

And then there are nations which have no coastline at all, or no appreciable one. These countries would have to obtain minerals from the sea by having the technology to challenge deeper ocean areas beyond the usual 200-mile limit and, in some fashion, mine the ocean floor, presumably renting marine harbor facilities from some accommodating coastal country. Or, these landlocked nations could, by treaty, get a share of what would be mined by other nations.

There has been, for many years, a proposal by landlocked countries to have just such a treaty enacted; but so far, it has not come to pass. These landlocked nations, in many cases, are also the smaller and less developed nations and given their lack of technology and, frequently, the lack of finances, their ability to in some way go out and mine the ocean floor is presently non-existent. It is the larger nations, with an extensive coastline, that are the potential players in the game of ocean mineral exploitation. It is these nations, also (as might be expected), which oppose any treaty stating they must share the wealth with the landlocked countries. The United States is one of the nations that has voted "no."

In the case of the United States, there is plenty of territory to explore, for in 1983 President Reagan proclaimed the ocean area from a line three miles offshore the coast of the United States (and its island territories) out 200 nautical miles to be the Exclusive Economic Zone. Presumably this is U.S. territory for mineral exploration and development. Especially with its islands in the Pacific — including the Hawaiian chain and Midway Island, Wake Island, Guam, and the Northern Mariana islands, American Samoa, and Howland and Baker islands — this gives the United States a huge amount of ocean floor to explore. The total area is 3.9 billion acres, compared with the total onshore United States territory of 2.3 billion acres.

How valuable this is remains to be seen, for the sea floor is known in detail only in very few places. Also, how economic it might be to recover minerals from this 3.9 billion acres is also unknown. It is indeed a vast frontier for the future.

Petroleum is a special situation, which is taken up in another chapter, so here we only briefly note that in recent years increasing amounts of petroleum are being recovered from beneath the sea, as land areas are becoming drilled up. For example, in Norway there is no on-shore production nor will there ever be, and in Great Britain there is very little. The North Sea is the oil province for both countries.

Even the countries around the oil-rich Persian Gulf, including Saudi Arabia, Kuwait, and Iran are, increasingly, moving offshore. The Gulf of Thailand and the South China Sea have more recently come into production, and Brunei gets most of its oil from offshore. Much Indonesian oil is also produced from beneath the sea. Australia, Nigeria, Trinidad, and Angola produce most of their oil offshore. Canada has moved offshore in the Newfoundland area, as there are few onshore prospects in eastern Canada. Canada is also moving, in a major way, north of the arctic coast. Mexico's big oil discoveries of recent years are offshore; and the United States, of course, has been in the Gulf of Mexico for many years, as well as offshore California.

In Alaska, the Cook Inlet area has been a petroleum producer for years; and such oil exploration frontiers as exist in Alaska are along the coast and offshore. Alaska has a land area of about 586,000 square miles; but given the geology of Alaska, it is exceedingly unlikely that any significant amount of oil will ever be found more than 50 miles inland.

But we are concerned with minerals other than petroleum in this chapter. For many years, shallow salt-water-covered sand and gravel shelves have been producing tin in Malaysia — the so-called Malaysian placer tin deposits. Elsewhere, along the coast of Africa similar types of deposits are mined for diamonds and for gold. However, in total value, these latter deposits are small.

There are various other materials of potential value on the sea floor. Notable among these are the very large deposits of manganese nodules, which lie on the mid-Pacific sea floor, between the two great industrial nations, Japan and the United States. These nodules are abundant in some places, to the extent of a pound and a quarter per square foot; and this is over many square miles. Although called manganese nodules, because this is the predomi-

184

nant metal in them, the nodules contain other valuable metals also and generally have a content of about 23 percent manganese, 6 percent iron, as much as 1.6 percent nickel, and lesser amounts of cobalt.

If these deposits were on land, they would definitely be an ore deposit. However, "ore" is not a geological term but an economic one, and implies a deposit from which a metal or metals can be extracted at a profit. Whether or not the manganese nodules can be recovered for their metal content — at a profit — is still uncertain. However, the quantities of metal there are exceedingly large, running into the billions of pounds. Also, an intriguing fact is that these nodules are continually forming today, with the result that even with extensive mining operations it is doubtful that as many nodules could be mined as are formed in a given year. Thus, we have the interesting situation where a mineral deposit gets larger, even as it is being mined. There aren't many mines like that!

Some experimental equipment has been built to recover these nodules. Results have been mixed. Also, the question has come up as to who owns these ocean minerals, as they are beyond the claimed territorial rights of any nation. As the Law of the Sea Treaty has not been endorsed by all countries (non-signers include the United States, West Germany, Great Britain, and Japan), it seems, for the moment at least, that the "law of capture" prevails — that is, whoever can *get* the nodules can have them.

There are other deposits on the sea floor, including phosphorite nodules. These exist in substantial amounts off both coasts of the United States, the west coast of Central and South America, in waters adjacent to Argentina, Japan, and South Africa, and off the central and northern coasts of New South Wales (Australia). These Australian deposits are fairly thin and, so far, poorly defined; but some are locally known to have a quality almost equal to those of onshore deposits, with a P_2O_5 content reaching 29 percent.

Recently, considerable interest has been shown in the now-forming brines around rift zones in the ocean floor. At these places, hot waters coming from the Earth's interior reach the ocean floor and precipitate a variety of metals. At the bottom of the Red Sea, there are such rift zones, in three distinct basins. Here metalliferous muds are relatively rich in certain metals, principally lead, zinc, copper, silver, and gold. Cores show that in some areas the muds are as thick as 300 feet. The concentrations here are from 1,000 to 50,000 times greater than in ordinary sea water. An analysis of some better deposits shows a content (by weight) of 0.12 percent manganese, 0.16 percent lead, 0.70 percent copper, and 2.06

percent zinc. However, they are still unable to compete economically with land-based deposits; but may become economic some time in the future.

Mineral crusts are also formed by rising, hot, mineral-laden waters; and these have been observed in a number of places, where "chimneys" of sorts are built up by what is called by oceanographers "black smoke," which actually is mineral-charged hot water. Such occurrences are known along the Gorda Ridge, off the northern California and southern Oregon coast, as well as a number of other places in the ocean floors. Some of these areas, including the California-Oregon occurrences, have had some preliminary mapping done on them.

Ocean water itself is a tremendous storehouse of elements. In fact, most of the elements known occur in sea water, in greater or lesser amounts — mostly lesser. Gold, for example, occurs in sea water in the amount of about 37.5 pounds per cubic mile, but this concentration is only 0.0005 percent. The cost in energy to process that much sea water, to produce so little gold, is inordinate. The same applies to virtually all other elements in sea water. The concentrations are so low that it does not pay to obtain them from this source.

There are, however, a couple of exceptions. Both bromine and magnesium have been obtained commercially from sea water. And, of course, in many countries, common salt is obtained from sea water by allowing the water to flood into shallow salt pans at high tide. Then the pans are diked-off from the sea and the water allowed to evaporate. Mexico, Thailand, and a number of other countries obtain common salt in this fashion.

What role all of these marine deposits — phosphorite and manganese nodules, hot brine muds, mineral crusts, and placers — will play in the future, for various nations, is difficult to assess. However, except for the richer placers being dredged now in relatively shallow waters, all of these materials can be recovered only at a very high cost in energy, so once again energy is the critical resource. In the case of sea water, as already noted, the amount of energy required to obtain various elements by processing the huge amounts of sea water that would be involved, is such that, except for a very few, the recovery of these elements from this source is unlikely for the foreseeable future. Such elements as may be recovered, and the metal deposits on the sea floor which can be obtained, will be recovered by a combination of technology and energy. So, it is energy, combined with the educated human mind, which will have to do the job, when and if it is done. For the immediate and reasonably near future, except for petroleum, the mineral resources of the sea will not be an appreciable factor in the futures of nations.

Chapter 14

Mining the Water Mineral

By almost any definition, water is a mineral, and if there is any one mineral which can be called "critical" or essential, water is it. Every living thing depends on water for its life. And water enters into our daily life in myriad ways. It is estimated that it takes about 40 gallons of water to put one egg on the table, 3600 gallons of water to grow a bushel of corn, 150 gallons to produce a loaf of bread, 375 gallons for five pounds of flour, and about 2,500 gallons to make one pound of beef. Your Sunday paper has taken about 280 gallons of water to produce; and one automobile, in all the processing of the raw materials that went into its manufacture, took about 100,000 gallons of water. And then there are all the water usages around the home.

The concern of this chapter is groundwater — water which occurs in the Earth at relatively shallow depths and which is commonly recovered for use by means of wells. According to the U.S. Geological Survey, the United States uses about 89 billion gallons of water a day, and of this, 20 percent is groundwater.

In considering groundwater resources, it should be noted that in some instances groundwater is replenished quite rapidly, whereas in other circumstances it may be depleted to the point where it could require decades or even centuries to recover. In some cases, it seems clear that the water-bearing stratum, called the aquifer, has been severely injured and never again will be as productive as it once was. Thus, in this discussion, we have a mineral which must be considered in these several contexts. If the groundwater of an area is being used faster than it is being replenished, it is sometimes said to be mined, even as other minerals are mined and therefore permanently gone. In the case of groundwater, however, it can be mined — that is, drawn out faster than it is recharged; but if the groundwater is brought under proper management, it can recover and be a permanent supply. However, in many areas, groundwater is indeed being mined.

Groundwater is a very important mineral in many parts of the world. In desert areas, with no permanent surface waters, groundwater is the only source of water available the year around. In semi-arid regions, groundwater is an important supplement to the less than adequate rainfall and allows crops to be grown where they otherwise could not survive. Over one-third of the irrigated land in the United States is watered by groundwater. More than 90 percent of people in rural America use groundwater for their domestic purposes. The windmill, combined with wells drilled in more remote areas of western United States, have kept ranchers in business, as cattle have been watered which otherwise, without the wells, could not have survived. About 20 of the larger cities of the United States get most (if not all) of their water from wells; and in twelve states, groundwater provides more than half of the total pubic water supplies.

About 97.2 percent of all the world's water is located in the salty oceans. Of the other 2.8 percent, all but less than one-half of one percent is tied up in ice caps and glaciers, is very deep in the Earth, or is in the form of atmospheric moisture. The total amount of usable water is only about .003 percent of the total water supply on Earth. The obvious water is that which we see on the surface, in the form of lakes and streams; but water in the ground — groundwater — makes up 97 percent of all fresh water in the United States. We shall not be concerned with the *surface* waters, the analysis of which would be a very large task in itself. But we will discuss *groundwater* that, in some respects, is like oil, in that generally one has to drill for it; and although some groundwater, unlike oil, is a renewable resource, other groundwater is so slow to recharge as to be in effect a one-time crop, even as is oil and other minerals.

Nearly 100 percent of the groundwater is rainwater or melting snow, which seeps into the ground and is stored for a greater or lesser length of time; and then may emerge as springs or, seeps, or be taken out of the ground by artificial means, through wells.

As the human race spread out over the globe, it invaded areas which are relatively arid. Water is needed. There are two ways to obtain it: one is by canals or pipelines, from distant water sources; and the other is by drilling wells. Pumping these wells may or may not be "mining" the water. If the rate of withdrawal is greater than the rate of recharge, it is a water mining situation. For many areas more recently, the waters are being used up faster than they are being replenished. The water which is being pumped out of the aquifer may have been in that aquifer for only a short time or it may have been accumulating for many hundreds, if not thousands, of years.

If an area, such as some of the desert regions of the Middle East and North Africa, has not had enough rain with which to grow crops, then, unless there are some special circumstances of recharge for an aquifer, from some distant, better watered area, the water taken out of the ground by wells to grow crops there is a depleting resource, because the rains clearly are not enough to replace what is taken out. If there were, then there would be no need to take water from the ground to grow the crops, unless there is a seasonal pattern of precipitation in the cold (non-growing) portion of the year. In the desert areas, this is generally not the case; and, therefore, using groundwater to grow crops where they cannot otherwise be grown, is a situation which can only be maintained for a limited time. The early result of doing this is that the water table drops. As this has happened in some areas, the obvious solution has been to deepen the wells, or drill into a deeper aquifer. But this has a problem also, in that the deeper waters are those which have generally been in the ground longer; and, also, as one goes deeper, the temperature of the Earth increases, so these deeper waters are also warmer waters. And as these deeper, warmer waters have been in the ground a long time, they tend to pick up more dissolved mineral material. Using these waters may ultimately make the irrigated lands salty and unusable. Finally, it costs more to pump water from deeper levels. Desert well-water agriculture has a number of problems.

The United States has been blessed with some exceptional underground water supplies. The glacial drift of the northern states is an easily-tapped, shallow, groundwater resource, which tends to be renewed each year with the rains and melting snows. And there are some deeper aquifers. The famous Dakota Sandstone, of South Dakota, at one time sustained flowing wells all the way east to the Minnesota border; but as more and more wells were drilled, this aquifer has had its pressure reduced substantially and many once-flowing wells flow no more.

The extensive Ogallala Formation, a huge underground lake, in effect, lies beneath parts of northern Texas, Oklahoma, New Mexico, Kansas, Colorado, Nebraska, Wyoming, and South Dakota, and is the principal geological unit in the High Plains aquifer, which underlies 174,000 square miles. This aquifer has a maximum saturated thickness of about 1,000 feet, an average thickness of about 200 feet, and, according to the U.S. Geological Survey, contains about 3.25 billion acre-feet of drainable water.

Recent count shows that more than 170,000 wells have been drilled into this aquifer for the irrigation of about 13 million acres of land. This acreage produces about 15 percent of the nation's total production of corn, wheat, cotton, and sorghum, and about 38 percent of the livestock. Estimates are that about 24 million acre-feet (an acre-foot equals one foot of water, over one acre, which is 325,851 gallons) are being taken out each year and that the recharge is only about three million acre feet. In places, the water level in the Ogallala aquifer has dropped more than 120 feet. The U.S. Geological Survey estimates that since this aquifer was first tapped, the volume of water in storage has decreased about 166 million acre-feet up to 1980. Computer models project a continuing net loss in water.

Clearly this situation cannot continue very far into the future. To solve the problem to some extent, there is a long-standing proposal to bring water in from as far away as the Mississippi River; but this idea has run into opposition from Arkansas and Louisiana residents, who do not want the Mississippi River flow reduced, especially after the severe navigational problems which occurred on the Mississippi during the summer drought of 1988. And Texas taxpayers themselves have turned down water financing proposals in the belief that the cost of nearly $4 billion was not worth it. In the meantime, the underground water, like the oil in Texas, is being depleted.

The same situation, to some extent, exists in the upper Great Plains area, where South Dakota, which now gets water from the Missouri River, (partly in an effort to preserve the Dakota Sandstone aquifer), is faced with demands from Nebraska, Iowa, and Missouri, who want to take more of the Missouri River water, as their groundwater supplies drop. The Missouri River in South Dakota is not a very large stream, so even this supply is limited.

These problems arise because an agricultural system is set up based on a resource which is depletable — in this case, ground-water which is not recharged as fast as it has been taken out, to support this agricultural economy. The water is being mined. It may be that the situation will ultimately be very much like that of other mining areas, where the ore has been taken out and ghost towns or greatly reduced populations remain.

For a number of years, the groundwater situation in Arizona has become increasingly critical. Each year it is estimated that Arizona uses 2½ million acre feet more of groundwater than is replenished by natural means.

Abroad, problems of groundwater supply in the wells of Ethiopia have been critical for years. Elsewhere, along with the political problems in the West Bank area of northern Israel, is the fact that the hills there form the water recharge area which is the major source of water for Israel's underground water reservoirs. The rainwater which soaks into the hills of the West Bank area spreads down into aquifers, which are used by most of the people in the central portion of Israel. This amounts to about one-third of the total water supply for that country. The increased Arab settlements in the West Bank area, with their correspondingly larger demands on both the surface and underground waters of the area, is causing water supply problems in the central area of Israel and will continue to do so and probably only become more acute, if present trends persist. Again, the problem is that the water is being withdrawn faster than it is being renewed.

In addition to the matter of lowering the water table, when groundwater is used faster than it is recharged — a trend which, if not checked, is clearly fatal to the area — there can be, and in many places is, another problem. The difficulty is that when fluids — oil and water — are taken out of the ground, the ground may subside, if the strata are not highly indurated — that is, fairly rigid, or the aquifer's materials are so well sorted, as, for example, in the St. Peter Sandstone, that further compaction cannot take place. Subsidence of oil fields in the Long Beach area of California has been observed for many years. Due to excessive withdrawal of water in the San Joaquin Valley of California, the surface of the ground has dropped as much as 15 feet, this extreme situation being in the Sacramento delta area. After the aquifer collapses in this fashion, it will not expand again from recharge. It has suffered permanent and irreparable damage.

In the San Joaquin Valley, about 30 percent of that area is subsiding, resulting from a drawdown of water levels, especially on the west side and in the southern end of the valley, (of as much as 400 to 500 feet). The three principal areas of subsidence are the Los Banos-Kettleman City area, the Tulare-Wasco area, and the Arvin-Maricopa area. Land subsidence in these places ranges from about one-half foot to a foot and a half per year. Pumping of groundwater had to be curtailed to prevent further subsidence.

Land subsidence in the central part of the Santa Clara Valley of California has been going on for more than 50 years, being first noticed when a detailed series of re-levelling studies were done in 1932 to 1933. A level line, established there by the Coast and Geodetic Survey in 1912, showed about four feet of subsidence in the San Jose area. A subsequent study by the U.S. Geological Survey, published in 1988, showed that the San Jose land surface had subsided a total of about 13 feet. This large amount of subsidence since 1933 is presumably due to the rapid and intense development of the San Jose area since 1933. It was concluded that the principal cause was the continual excessive pumping of groundwater. This same study showed that the artesian water level had dropped as much as 200 feet, since 1916. Now, however, due to imports of surface water and a decrease in groundwater withdrawal, the water level has recovered 100 feet or more, but the land subsidence remains.

In the Las Vegas, Nevada valley area, excessive pumping of groundwater there has caused a subsidence of as much as five feet. Some 400 square miles of this valley now are showing land subsidence, due to groundwater withdrawal. Visible displays of this situation can be seen by the well casings and their wellheads, which now stand as much as four feet above the ground level at which they were originally placed. It is estimated, also, that excessive pumping has reduced groundwater levels as much as 180 feet around the Las Vegas Valley Water District well field.

Again, it should be noted that once these aquifers have been damaged by subsidence, their ability to hold and to produce water is permanently impaired. Any agricultural economy built up on this groundwater resource, which cannot be renewed to what it was once was, will either have to contract or find some other source of water. Texas built a great economy on oil and now it is having to adjust to reduced oil production. California's San Joaquin Valley has been one of the most productive and most valuable pieces of agricultural land in the world. With the mineral resource, groundwater, on which this economy was partly built, now being partially lost, the economy of the valley will have to adjust accordingly. Surface water brought in from considerable distances, by canals, is one solution being used.

In Arizona, large land areas in the southern part of the state have been slowly subsiding. Great quantities of groundwater have been pumped out of this region since 1900; indeed, more has been withdrawn than has been replaced by natural recharge. Subsidence was first noted in 1948, near Elroy, in south-central Arizona. As of 1977, a re-survey showed that since 1948 the ground subsidence has been more than 12 feet in places, with the maximum annual subsidence rate being about 5½ inches.

In Arizona's Salt River Valley, where fields have been irrigated by groundwater for many years, the water table has dropped more than 300 feet in places. This loss of water in the ground has caused subsidence of the surface in a number of areas. The Central Arizona Project aqueduct, which brings in Colorado River water to Arizona, had to be especially constructed to compensate for continual land subsidence, near Apache Junction. Elsewhere, in Paradise Valley, which is a residential complex northwest of Phoenix, a 400-foot long fissure opened as an apparent result of groundwater withdrawal. Many more fissures are now known, with hundreds occurring in a number of the basins in parts of Cochise, Maricopa, Pima, and Pinal counties — all areas of groundwater withdrawals for agriculture.

The effects of these land subsidences are becoming costly. The flow of sewers may be impeded and, in some cases, even reversed. Bridges, tunnels, railroad lines, power lines, and highways are all adversely affected. Cracks in the foundations of buildings occur. A railroad derailment was caused by land subsidence and shifting of the rails. So, in total the mining of groundwater has become, in places, a substantial economic liability.

And there is a still further negative effect of mining groundwater. Along the coastal plain of southeastern United States, several excellent freshwater aquifers dip gently toward (and ultimately into) the sea, as, for example, in Virginia. But because they are open-ended in salt water, keeping the salt water out depends on a steady flow of fresh water from the higher level of the land, and toward the ocean. If fresh water is drawn out excessively, the hydraulic pressure of the fresh water is reduced; and there is a landward invasion of salt water into the aquifer. This has been a problem in Norfolk, Virginia, and also occurs elsewhere in the coastal plain.

In regard to a fresh water-salt water relationship, a particularly delicate situation exists in some of the Pacific islands, where lenses of fresh water sit on top of salt water, in the very porous coralline islands. If the fresh water is withdrawn beyond its recharge rate, the salt water rises and contaminates the wells. Reversing this situation takes considerable time.

In summary, like oil or hard minerals, groundwater (in places) can be a one-crop situation. If the rate of recharge is such that replacing the withdrawn water takes years or even centuries, groundwater is then, for all practical purposes, a crop which can be harvested only once. If excessive groundwater withdrawal results in collapse of the aquifer and land subsidence, then a greater or lesser amount that aquifer's groundwater will become a permanently mined out condition. If a region is developed on the basis of overdrawing the groundwater, this can continue for a time; but ultimately the development must stop and then contract, unless other water supplies are brought in.

From time to time, especially in the Middle East, groundwater irrigation projects are proposed and a number are already in existence. It is doubtful that these will be viable very far into the future; and as this resource diminishes, serious adjustments will have to be made.

Statements are commonly made to the effect that beneath the desert sands there is a huge underground lake. The lake, of course, is simply a sand or gravel deposit or a porous rock of some sort, which is saturated with water; but the water probably took decades, if not thousands of years, to accumulate and also it may be moderately to highly mineralized. In any case, withdrawing this water at a rate to sustain substantial agriculture, over any long period of time, is probably not possible. It is an interim solution, at best. It is much like a metal deposit, with a finite life. The groundwater is being mined.

In the case of the United States, a virgin land was taken over and in many places the groundwater heritage from ages past is now being used in a manner which clearly cannot be sustained. In other parts of the world, increasing populations are drawing more and more heavily on groundwater supplies. The efficient management of these resources as a continually renewable resource is vital; but unfortunately this is not being done in many, and perhaps in most, areas. And, as we have noted, in some places irreparable damage has already been done to some aquifers.

Groundwater problems usually are not of a sufficiently large scale to greatly affect a nation as a whole, although in smaller countries and those in more arid regions, as, for example, in parts of Africa, the effect of reduced or destroyed groundwater supplies can be significant. Nevertheless, even in larger countries, such as the United States, where the Great Plains are being affected, the long-term, adverse results of mining groundwater are likely to be both substantial and felt on a national scale.

Chapter 15

Minerals and Medicine

Most of this book is concerned with the economic and military health of nations, with respect to the distribution and availability of mineral resources. But there is another facet of mineral distribution — which affects the *physical* health of individuals and of the domestic animals on which humans depend. The weathering of bedrock, or in some areas, the glacial drift, or otherwise transported Earth materials, to form soils, ultimately determines what elements are in the soil that can be used as plant nutrients and will then be eaten by animals and by humans in the area.

As the bedrock or underlying transported rock materials can be quite different, from place to place, and in the past (and even at the present time) as many populations derive their sustenance from relatively small and local areas, there is a good possibility that vital trace elements that are needed for proper nutrition can be missing from the diet of a population in a given locality. Actually, this might apply to regions as large as some countries.

The study of trace elements in human, as well as animal and plant nutrition, has turned up some interesting facts with regard to distribution of trace elements and their relationship to diseases, in various parts of the world. For example, there is evidence from the geographic distribution of thyroid disease, hypertension, arteriosclerosis, cancer, tooth decay, and from several diseases of animals, that there is a definite relationship between geochemistry — that is, the chemistry of the Earth in those places — and these medical conditions.

The United States imports almost 100 percent of its chromium; chromium is a relatively rare element in the United States. There are small deposits in Montana and Oregon and in a few other places; but for the most part, both North America and South America are markedly deficient in chromium. Is this chromium deficiency identifiable in humans, and is it significant in terms of health? The answer appears to be "yes" to both questions. Near-Easterners have about 4.4 times as much chromium in their bodies as do Americans; and Orientals have five times as much. In studying death rates and causes, it was found that chromium in the aorta was too low to be detected in every person dying of coronary artery disease, one aspect of arteriosclerosis (Schroeder, 1973).

Chromium's presence has been found to be important in reducing cholesterol levels, which apparently accounts for its absence in people dying of the coronary artery problems, just cited.

Selenium is an element which seems to both cause and cure a variety of human ailments. A study of 45,000 Chinese, with regard to the occurrence of Keshan Disease, which is a form of heart disease wherein the heart is enlarged, there is low blood pressure, and a fast pulse, a high death rate was found to be clearly related, geographically, to the amount of selenium in the soil. Low selenium areas had a high rate of Keshan Disease. In the United States, an area along the coastal plain of Georgia and the Carolinas has come to be termed "the stroke belt," and also has a higher than normal incidence of heart disease. This area, too, as in China, is low in selenium. Although the studies are not yet complete, it appears that death rates from a variety of cancers are less in areas of the United States where local crops take in larger amounts of selenium. The lack of selenium is one of the earliest cases of mineral deficiency observed, for Marco Polo described this during his travels in China, noting that hoofs of cattle went bad and were dropping off. He did not know, of course, that it was due to lack of selenium but did relate the disease to a particular region.

Selenium poisoning can also occur from an excess of this element. It has been observed for a long time, that in some parts of South Dakota cattle may be victims of selenium poisoning, due to excesses of the metal in the local soil. This has resulted in what has been called "alkali disease" or "blind staggers."

It also became well established, over the years, that Wyoming ranchers cannot successfully graze sheep on a particular geological stratum (specifically the Niobrara Chalk, or on certain areas of the Mesaverde Sandstone) in the spring, especially in wet years, due to selenium poisoning. The sheep would lose muscular control and frequently died. Selenium poisoning is quite widespread around the world, being known in Canada, Mexico, England, Wales, Colombia, Argentina, Israel, Zaire, Nigeria, Kenya, India, South Africa, Australia, and Japan.

An extreme problem of selenium poisoning developed in the northwestern part of the San Joaquin Valley, when the San Louis Drain was not completed into San Francisco Bay, but was simply dumped into the Kesterson Wildlife Refuge wetlands area. These waste irrigation waters were high in selenium; and as they accumulated and evaporated, the selenium content increased to the point of being highly toxic. Deformed birds of many species resulted, to the extent that men were hired to go through the refuge and discharge firearms, to prevent waterfowl from staying there and nesting. So far, the toxic effects are confined to the wildlife of the refuge; but there is concern that the selenium will ultimately get into the groundwater of the area, which is used for human consumption.

In late 1988, a general selenium poisoning warning was published by the Sacramento Bee (California), as a result of that paper's on-going investigations which found selenium contamination in the marshes, lakes, and streams, and the wildlife of those areas in Wyoming, Colorado, Utah, Montana, and Nevada, as well as in California. Fish and game from these areas contained excessive amounts of selenium; specifically 81 percent of the trout, carp, perch, catfish, and goose eggs collected by that paper, in its study throughout the West, exceeded the 200 microgram safety limit; and 67 percent were over the 500 level of toxic effect. The samples averaged 974 micrograms or nearly double the level at which poisoning symptoms begin to appear in healthy human adults.

The most notable case of excessive selenium was in Sweitzer Lake, at the state recreational area of the same name, near Delta, Colorado, where the water tested 51 parts per billion of selenium, which is 10 times higher than the Environmental Protection Agency's limit for protecting freshwater fish species.

Products for human consumption were studied and half the foods tested, such as steak, liver, poultry, eggs, and vegetables from parts of Oregon, Montana, South Dakota, Nebraska, Wyoming, and Colorado, were found to exceed the safe level of 200 micrograms of selenium. The true magnitude of this situation, in western United States, has yet to be determined; but the clues are already there to indicate that the problem could be large.

The Kesterson problem and what may be a more general western United States problem is one of excess selenium; but as has been noted, a deficiency can also be hazardous. In British Columbia, cattle are injected with selenium to cure muscular dystrophy. In New Zealand, where there is no natural selenium in the soils, selenium has to be included in the fertilizer, to prevent decimation of the flocks of sheep on which New Zealand is dependent for substantial income. Selenium, it was found, was a catalyst to the enzymes which, combined with vitamin E, operated to insure the production of viable offspring.

Another example, relating to the presence or absence of a particular element on livestock, is known as "grass tetany," which has been reported from Ohio. This results from the lack of sufficient magnesium in the diet to maintain normal magnesium levels in the serum of the blood stream. Cattle do not have a readily usable magnesium reserve, and the symptoms of grass tetany can develop quite rapidly (within 24 to 48 hours). The cattle may die within a few hours. Female cattle are particularly subject to this ailment, shortly after giving birth to their calves. It was found that this affliction was chiefly in a 26-county area, where the soils are primarily derived from sandstones and shales, both of which (in that area) are deficient in magnesium. The apparent solution to this problem was to increase the magnesium content of the soil by applying high magnesium content limestone (dolomite). This fortunately is also a long term solution, as the dolomite weathers slowly and gradually releases magnesium to the soil over a period of as much as 10 years. The immediate solution to the problem was made by dusting pastures with magnesium oxide.

On a healthier note, it had long observed (without knowing the cause) that certain regions of the Serengeti plains of East Africa attracted a particularly large number and variety of animals. Botany Professor Samuel McNaughton, of Syracuse University, measured the levels of 19 minerals in some of the Serengeti grasses and found that the grasses in regions which the animals favored had much higher levels of minerals, particularly phosphorus, magnesium, and sodium, than did the grasses in adjacent areas.

Fluoride, like selenium and many other elements, can be both beneficial or toxic, depending on the amount. Found naturally in some drinking waters, in moderate amounts (in the vicinity of one part per million) fluoride appears to be a substantial deterrent to tooth decay. On the other hand, excessive quantities of fluoride in drinking water causes undesirable bone changes, as well as mottling of the teeth. The function of fluoride in both teeth and bones is to help retain calcium. Where water is deficient in fluoride, as is the case in certain areas of North Dakota, there is an increase in osteoporosis, which is a decreased strength of bones, due to leaching of calcium, an ailment which tends to affect older persons, especially women. There also is evidence that taking calcium fluoride pills can decrease the incidence of hardening of the aorta. Deficiencies of magnesium, calcium, and lithium are correlated with an increase in cardiovascular troubles in general. Hard water appears to reduce these problems.

A number of years ago, the term "salt sick" was applied to animals which exhibited weakness, failed to fatten, and in general were anemic. It was found that cattle developed this trouble when they foraged only on grass grown on certain, light-colored sandy soils and some peat soils. Examination of the soils showed they were deficient in both iron and copper, compared with other soils. When iron, copper, and cobalt were added to the soils, the problem disappeared. Children in the areas of these "salt-sick" type soils (which were in Florida) were also found to be anemic; and treatment of the children with iron greatly improved their condition, usually in from four to six weeks. In Florida now, areas of deficiency in mineral micro-nutrients are fast disappearing, because fertilizers, with mineral additives especially tailored to the particular soil problems, are now being used.

Every living cell has to have, among other things, two vital elements — potassium and phosphorus. Phosphorus tends to get locked away in mineral form quite easily, instead of freely circulating. Bones and teeth are calcium phosphate. Simply to build the bones of a billion Chinese takes a lot of phosphorus. Keeping phosphorus in circulation, so that all people can get their needed share, is somewhat of a problem already. It may be more of a problem in the future. Morocco has the world's largest phosphate deposits. Perhaps Morocco will become an especially critical area to the world economy, even as the oil-rich nations are now.

Soil differences in areas not far apart can result in appreciable health differences. It was noted in New Zealand that there were substantial differences in the amounts of tooth decay between children in the city of Napier and the nearby city of Hastings. Soil analysis from the areas adjacent to the cities showed considerable differences in trace elements in the vegetables grown in the two areas. Tooth decay was lower in the Napier area, where the soil was higher in molybdenum, aluminum, and titanium, and lower in manganese, barium, and strontium, compared with the soils of the Hastings area, where the vegetables grown there reflected these differences. The teeth of the Napier children were higher in molybdenum than those of the Hastings children. Also, in New Zealand it was found through a careful study made by the Medical Research Council of New Zealand, the U.S. Public Health Service, and the U.S. Navy that areas where there was a low fluoride content in the soil and the groundwater contained less than average calcium and boron, and where there was a low calcium/magnesium ratio, that the susceptibility to dental decay was relatively high.

Arsenic in drinking water is a fairly common problem in several parts of the world. In Oregon's southern Willamette Valley, near the village of Creswell, there are excessive amounts of arsenic in the groundwater used for local water supplies. The source of the arsenic is not precisely known but it probably comes from grains arsenopyrite in the aquifer. Arsenopyrite is a fairly common, arsenic-bearing mineral, in regions where basalt or other basic igneous rocks are abundant — which is the circumstance near Creswell. In Taiwan, drinking water from wells produced arsenic poisoning in hundreds of persons. The arsenic exceeded 70 parts per million and many of the affected Chinese developed skin cancers, along with the general arsenic poisoning.

Lead is a metal that has been in use for more than 2000 years. The Romans made their water pipes from lead, some of which are still very well preserved, as for example, in the ruins of Pompeii. It has been suggested — with a considerable basis in fact — that the Romans may have suffered, as a population, from lead poisoning, because of this use of lead in their water supply systems. The U.S. Geological Survey, in analyzing municipal water supplies of the one hundred largest cities in the United States, looked for 23 trace elements, and of these, 16 were found in sizable quantities. Of these, five are essential and 10 are biologically inert, and only one, which was lead, is toxic over a lifetime. The maximum concentration of lead which was found was 62 parts per billion, which exceeds the Federal lead safety standards, which is 50 parts per billion. Such lead as was found occurred in water supply systems where the water was soft and acid, and which included some lead pipes. The lead was not an initial constituent of groundwater, all of which suggested that the Romans might have indeed picked up lead from their water systems.

Recently, especially in the United States, considerable concern has been expressed regarding radon gas, as a possible health hazard. This gas is derived from a series of radioactive decays which start with uranium. Actually, radon gas itself has a very short life; and it is the subsequent decay products which cause cancer. These accumulate in houses and other closed structures. The occurrence of radon gas and its derivatives can definitely be related to the geology of the area. Uranium in particular is found in certain black shales, such as underlie parts of Kansas and some other Midwest regions, and in granitic rocks. Granitic rocks are the original source of the uranium and upon weathering, it is transported in oxidized form and then immobilized in the reducing environments of organic-rich muds, which upon compaction, are black shales. Both black shale and granitic bedrock areas usually show a higher than average radon gas presence. Certain other rocks may also produce radon; but, in any case, the geological associations of radon are becoming well documented. Geological mapping is proceeding in several areas to narrow down the sources.

Heart disease has been studied, perhaps more thoroughly than any other human ailment; and here there seems to be at least one relatively conclusive relationship to minerals. In England,

Wales, the United States, Sweden, and Japan, numerous studies have shown that areas which have relatively soft water (that is, water with few dissolved minerals) have a higher than average death rate from heart disease. Hard water areas, for example where water is drawn from wells in limestone areas, appear to have populations that have distinctly less heart disease (which includes both cardio-vascular disease and coronary heart disease) than areas where there are fewer minerals in the water.

In Japan, a relatively small geographical area, this relation-ship has been defined quite well. Northeastern Japan has an abundance of sulfur-rich volcanic rocks and it was found that rivers here carried relatively soft water, as the high sulfate produced an acid water, which was soft water. These areas proved to be places where the death rate from apoplexy (death due to rupture of a blood vessel) was very substantially higher than in areas to the south, where the river waters were relatively hard. Studies in various regions of the United States show this same relationship — a negative correlation between heart disease and the hardness of the local waters. Soft water may be fine for your car battery, and for washing your hair, but it is bad for your teeth and heart. It should be noted, however, that the precise factors involved in this correla-tion between heart disease and hardness (or softness) of water are still undetermined. That is, just exactly what hard water does to inhibit heart disease is still not known with any certainty; the correlation simply exists.

Other relationships between heart disease and the geology of an area have been noted. For example, in Georgia, nine northern counties have a notably low rate of heart disease; whereas in south-central Georgia, the death rate from heart disease is appre-ciably higher. A detailed geochemical analysis of the soils and the plants in the two areas disclosed that in the northern, low-death-rate area, manganese, vanadium, copper, chromium, and iron are more abundant than in the high-death-rate area. These elements are among those trace elements which are known to have an advantageous effect on heart disease.

The scourge of mankind, cancer, also appears to have some definite geological associations; but again, as in the case of heart disease, exactly what causes many cancers is unknown. In West Devon, England, it seems clear that a high incidence of cancer is related to groundwater obtained from a distinctive rock in the area, from which very highly mineralized water is obtained.

The geography of esophageal cancer, near the Caspian Sea, in Iran, also shows a distinctive pattern, which is related to soil types. The soils of the eastern portion of this area are saline, being a relatively dry area. Going westward, the amount of rainfall increases and the salinity of the soils markedly decreases; and along with this trend is a marked decrease in esophageal cancer. As a follow-up to this study, it was found that other areas of the world which have a high rate of esophageal cancer also have much the same nutrition, including mineral content of the soils, as does the population of the high-risk areas in Iran. One such area was found in Puerto Rico. Another area of excessive esophageal cancer that had a climatic environment, which in turn affected the mineral environment (high salinity), similar to the high-cancer-incidence area in Iran, was found in the Turkoman (USSR) semi-desert region.

Very locally, the effect of certain minerals in an environment can be quite striking. Women living in a particular silver mining area in Bohemia used to outlive three or more husbands who worked in the silver mines. These silver mines are now a major source for uranium. The miners did not know it at the time; but they were being subject to excessive radiation while working in these mines, and also were subject to lung problems from silica dust resulting in silicosis; and the fine silica dust also appears to have contained uranium contamination.

On a broader scale, the renewing of mineral resources for an agricultural area is important. This is usually done by humans through the application of fertilizers, such as ammonia and phosphate; but in many parts of the world, these are not available or cost more than the local farmers can afford. Lands which are continually leached by heavy rainfall eventually become relatively infertile, the classic example being the Amazon Basin. Immediately along the rivers which bring in nutrients from distant mountains and locally flood across the adjacent ground, the land can be relatively productive; but on the higher areas, away from the river flooding, the lands may have been continually leached, literally for millions of years, and virtually no usable mineral nutrients remain. This has given rise in the past to the "slash and burn" type of agriculture. More recently, attempts have been made to clear the jungle and plant conventional crops; but generally this has met with failure. The leached, residual, red clays, when cleared of their normal jungle cover, tend to bake like a brick under the tropical sun and become essentially nonproductive.

In some regions, however, nature does renew these leached soils. On the islands of Indonesia, particularly Java, which has one of the highest population densities in the world, volcanic activity frequently dusts the landscape with a fresh layer of mineral material, which quickly weathers down in that warm, moist climate to release new minerals, producing a fertile soil. Without these volcanic eruptions, it is doubtful that Java, in particular, could sustain its present population; and in any event, there would surely be some trace element deficiencies, without the mineral renewing effects of the ash falls. By the same token, the flanks of Mount Vesuvius have seen several very destructive times; but immediately after each eruption, the people there have been quick to move back up the slopes and replant the vineyards and other crops, for the newly deposited mineral-rich ash soon breaks down to produce bumper crops.

The 1980 explosion of Mount St. Helens, in the State of Washington, dumped large amounts of ash over central and eastern Washington's great wheat growing area, and into northern Idaho and beyond. The immediate effect of this was negative; but the farmers in this region can now look forward (given normal moisture conditions) to bumper crops, for many years to come, as a result of this extensive and free natural fertilizer deposit.

In a different way, nature has recently (geologically speaking) renewed the fertility of a substantial portion of the Upper Midwest area of the United States, including that very fertile area called Iowa. Iowa and adjacent parts of America's crop heartland produce wonderfully rich crops, full of the nutrients needed for good health. The roots of these crops are in glacial drift — a deposit of varied materials, hauled down from the north by glaciers in the quite recent geological past.

These fresh rock materials were derived, in large part, from granitic areas which have an abundance of the minerals needed for good health. After deposition, the glacier left and weathering took over, to make these elements available to plant life. Also, the cold winters preserve the rich black humus from the decayed vegetation of the previous year. This was the inheritance from nature that the first settlers found and from which people in the Upper Midwest still benefit, and will continue to benefit for generations to come. It is a mineral heritage brought down, duty-free, from Canada, by the ice sheets.

One can readily see, however, that it would have been a vastly different story without the glaciers, for the road cuts and the quarries of this area show that beneath the relatively thin cover of glacier drift lie hundreds to thousands of feet of leached marine clays, laid down in ancient seas which covered this part of the United States some 300 to 400 million years ago. Fossil corals, cephalopods, bryozoans, trilobites, and many other forms of marine life lie here beneath the glacial debris in Iowa, in relatively infertile clays that would be difficult material, indeed, in which to grow 150 or more bushels of corn to the acre. Also, such corn as might be grown would be deficient in a number of vital trace minerals.

The apparent importance of these mineral-rich, glacial soils to longevity was reported in a 157-page study prepared by a panel under the direction of Dr. Howard Hopps, Professor of Pathology at the University of Missouri. The study compared death rates of men ages 35-74, in two 100,000 square mile areas. One area was in the glaciated, Upper-Midwest, mineral-rich soil and groundwater area, and the other was in parts of the southeastern coastal area in Virginia, the Carolinas, Georgia, and central Alabama. This latter area has a meager supply of minerals in its drinking water and soil. The report stated that "for every 100 men in this age range who died in a given year in the Upper Midwest region, 200 died in the coastal area."

The panel found that cardiovascular diseases, primarily heart attacks and strokes, accounted for most of the differences in deaths between the two areas. Dr. Hopps said that the Upper Midwest was left rich in minerals and trace elements by the glaciers that "ground up the rocks and made minerals in them available." These minerals include iron, copper, manganese, fluoride, chromium, selenium, molybdenum, magnesium, zinc, iodine, cobalt, silicon, and vanadium, according to Dr. Hopps. In the southeast, Dr. Hopps noted that "the minerals have been leached out of the soil for millennia." He also noted that the differences were consistent, stating that "no county in the Minnesota part of the region, for example, was above average in deaths. It seemed to be an inescapable conclusion that a lot of people in the Upper Midwest must be living a lot longer." The study focused on white men, to rule out the possibility of different racial makeups in the regions, that might affect the results. The panelists said that apparently trace minerals in the soil and water do contribute to relative longevity for persons living in this area of glacially transported materials, as compared with other areas less fortunate in having fresh, new, rock materials from which to weather out vital elements into the soil.

However, a warning should be sounded. Already in places, due to careless plowing and soil conservation practices, the red and grey ancient marine clays are beginning to show through the tops of hills in Iowa and elsewhere, indicating that this rich glacial heritage from the recent geological past may be disappearing. Unlike volcanoes of other lands, the glaciers are not likely to reassert themselves very soon, to again refurbish the mineral content of the soils.

The presence of certain bacterial diseases can be related to soil types. The distribution of anthrax has been shown to be geologically controlled, to a considerable extent. It is a disease which develops in areas where the soil pH is more than 6, and the minimum temperature is 60 degrees. The soil conditions favorable for anthrax are associated with limestone terrains, alkaline alluvial soils, and clay soils where there is a hardpan layer. The occurrence of anthrax is rare on well-drained, sandy soils or shale soils.

Can minerals in the soil, or lack thereof, affect the health of a population as large as a nation? Perhaps. In the case of some of the recently established smaller island nations of the Pacific, in particular, some of the islands are simply atolls — that is, coral reefs. Corals are very selective in what they make their reefs from; the material is nearly pure calcium carbonate. If it were not for the fish which the inhabitants of these islands have incorporated into their diets in the past, their daily meals would have been markedly deficient in vital minerals, for soil developed from pure limestones is very poor in essential trace elements.

In general, lack of minerals would more likely be a problem in a small nation, where there is no great diversity of geology and therefore of minerals, as compared with a nation of larger area. But even on a larger scale, lands which have been leached by rainfall over thousands and, in some cases millions of years, such as many parts of the Amazon Basin of Brazil, are not nearly so fertile as the relatively fresh, glacially-derived soils in the Upper Midwest of the United States. Correspondingly, the populations of the two areas in all likelihood would be found to have different levels of health, the advantage clearly being with the population deriving its sustenance from the fresh, mineral-rich, glacial soils. And indeed, over the centuries, the vitality of one nation due to differences in bedrock and soil geology may well be a contributing factor to its destiny. The nations with the better minerals in their soils will be the nations with the healthier populations. People living off the glacial, drift-derived soils of Illinois, Minnesota, and Iowa will definitely be better nourished than those on the leached laterites of Brazil and parts of Africa.

Chapter 16

Strategic Minerals —
How Strategic Are They?

From time to time the press reports there is national concern about supplies of "strategic minerals." There is unrest in South Africa, source of much of the world's platinum. There is political instability in Zaire, source of much of the world's cobalt. How important are these minerals to the rest of the world, particularly the highly industrialized nations? And, in turn, how much of a power struggle may sooner or later occur over these materials?

A considerable difference of opinion exists concerning the importance of strategic minerals. Some say that these materials are of great and vital concern, and their free flow through international trade must be maintained at virtually any cost. Others contend there are adequate substitutes, and these, combined with government and private stockpiles of such materials, make the prospect of a cut-off, at any given time, of little consequence.

Before pursuing this matter further, the term "strategic mineral" should be defined. The U.S. Bureau of Mines advises me by letter, dated February 1, 1988 regarding "strategic" and "critical," that "There have been many attempts to define each term by a large number of individuals, however, the Bureau does not make a distinction between the two terms. The Defense Production Act of 1950 regards 'strategic' and 'critical' essentially as the same. Section 12 of the 1950 Defense Production Act states: For purposes of this Act: the term 'strategic and critical materials' means materials that (A) would be needed to supply the military, industrial and essential civilian needs of the United States during a national emergency, and (B) are not found or produced in the United States in sufficient quantities to meet such a need. The act goes on to designate energy as a strategic and critical material."

Under this rather broad definition, virtually all minerals are critical which are not produced in sufficient quantity in the United States to make the USA independent of foreign supplies; and virtually all mineral materials and energy minerals are needed to some degree, in industrial and military establishments. However, by what is a more narrow definition but actually a more common usage of the term "strategic" (or "critical"), there are certain metals which are listed, and oil and other materials are regarded separately. We shall, for the moment, employ the more common usage of the term "strategic," restricting it to metals, and to metals which are of special note, because few deposits of them exist in the world and, therefore, they are not readily available to most countries — metals which are important industrially and for which, in general, no adequate substitutes exist.

A common example of minerals which are in this category has frequently been given in the form of a list of metals which go into a jet airplane engine, a piece of machinery essential to both the military and civilian economies of the world:

5366 pounds of titanium

5204 pounds of nickel

1656 pounds of chromium

910 pounds of cobalt

720 pounds of aluminum

171 pounds of columbium

3 pounds of tantalum

Actually, just which minerals are critical and which are not depends on the country. Each nation has its own resources and its own external needs, and thus has its particular list of what might be called critical minerals. The vulnerability of the various countries and areas (European Economic Community) to being cut off from particular minerals differs rather widely with the commodity. In the minerals usually thought of as strategic, the USSR is by far the most self-sufficient, and the United States is intermediate.

There are also wide differences as to the importance of each mineral, and the degree to which it can be substituted. Gold, for example, has few critical uses and its absence would hardly paralyze a country. On the other hand, manganese is a key metal in the making of steel; and there appears to be no substitute. Tungsten also is a critical material for which, again, there is no apparent good replacement in quantities which would be needed. But bauxite, the chief ore of aluminum, is critical only in the sense that it is the ore of aluminum, from which the metal can most easily be recovered. Aluminum constitutes eight percent of the Earth's crust. There is no shortage of aluminum in the ground virtually anywhere; ordinary clay is an aluminum compound.

In the case of bauxite, nature has partially done the job of breaking the tight aluminum bond with other elements and thus reducing the amount of energy needed to recover the metal. If energy were cheap enough, obtaining aluminum from common clay would be no problem, for the world is loaded with clay.

The following table is a selected list of what are generally regarded as strategic minerals, and which are held by South Africa and the USSR. (Source: U.S. Bureau of Mines).

South Africa And USSR
Percentages of World's Reserves
of Selected Strategic Minerals

Mineral	S. Africa %	USSR %	Combined
Platinum group	86	13	99
Manganese	53	44	97
Vanadium	64	33	97
Chromium	95	1	96
Uranium	27	13 (est.)	40
Titanium	5	16	21

This tabulation is (currently) essentially correct; but it should be noted in the case of chromium that there are small deposits in a number of countries around the world, which have not been included in this tabulation. Also, recent discoveries of uranium in both Canada and Namibia, where there already are large deposits known, may change these figures somewhat. However, they are broadly accurate in showing that South Africa and the Soviet Union are, indeed, rather well endowed with some of the more critical metals.

Supplies of these materials are important, for, as the American Geological Institute has stated, "Without manganese, chromium, platinum, and cobalt, there can be no automobiles, no airplanes, no jet engines, no satellites, and no sophisticated weapons — not even home appliances."

As noted in the early part of this chapter, we have excluded oil from our "strategic minerals" list, because it is a rather special and large situation and is given separate treatment in chapter 10, *The Petroleum Interval.* But we briefly comment here that oil is certainly a strategic mineral for the countries which have none. And it is becoming so for other countries which have some oil, but as industrialization increases and/or oil supplies decrease, more and more oil must be imported. Such is the situation for the United States. Until about 1970, the United States was self-sufficient in oil; and it was not a strategic mineral, but it surely has become such since that time.

Oil in 1920 was not a strategic material, for in that year the United States produced nearly two-thirds of the entire world's oil; and no one had ever heard of Saudi Arabia, for the country of that name did not exist until 1932, and, of course, OPEC was also non-existent. Now the United States accounts for less than 20 percent of world oil production, and its reserves are dwindling rapidly. So what was a very abundant and non-strategic material in 1920, has now become in domestic short supply and can only become more scarce as time goes on. So oil, now, is entering the category of a strategic mineral for the United States.

Thus, the concept of what are strategic minerals, using the criterion of adequate supply within domestic borders, changes as exploitation and depletion of these materials proceeds, and industrial, military, and general civilian demands increase. On this basis, one can say that in what are currently the world's major industrial countries, the number of strategic minerals can only increase with time. In terms, therefore of nations' needs, and the sources of these materials, the balance of power will shift toward those countries and areas which have the critical raw materials. In considerable number, these are the lesser developed countries. Seeing this, some of these countries have tried to form mineral cartels, using OPEC as a model. A bauxite (aluminum ore) cartel was attempted at one time, with no success. To date, all these efforts to form raw materials cartels have been unsuccessful, with perhaps the only exception being the diamond cartel, consisting of a single company, De Beers, Ltd. "Metals" cartels may become successful in the future; but this appears unlikely, at least for the near term. And the oil cartel, OPEC, has had its problems, too, being only partially successful to date. This we treat in a subsequent chapter.

What we are struggling with in this matter of the distribution of strategic materials are what might be called "geological accidents" (although there are reasons behind all things geological, which, however, may at present not be completely understood), accounting for the fact that mineral resources and energy mineral resources are very unevenly distributed around the world. The great silver deposits of the globe are largely in the Western Hemisphere, near the west edge of both North and South America (Peru, Mexico, and the United States). Platinum, on the other hand, is chiefly found in just two places — South Africa and a relatively small area of the Soviet Union.

These diverse occurrences of minerals, combined with the present political boundaries, cause the problems of access to minerals by the world's economies. But the political boundaries and the uneven distribution of vital materials are facts of life not likely to disappear. The only peaceful solution apparent at this time is free trade.

There are, however, two factors which might appear to change this situation somewhat. One is that perhaps a technology can, in particular situations, devise substitutes for what are now regarded as absolutely essential strategic minerals. (In December, 1988, Ford Motor Company announced a substitute for platinum in automobile emission systems. However, no substitute is yet known for platinum in its many other vital uses). The second prospect is that changing lifestyles, and/or new and different social and economic systems, will reduce or eliminate the need for a particular material. However, given the desire of people to maintain (and, for many, hopefully even increase) their standard of living, this latter suggested solution seems unlikely. Demand will continue to grow for minerals.

It is probable that the strategic materials each country needs today will continue to be strategic; and as use increases against finite supplies and the higher grade materials are depleted, leaving the more expensive lower grade deposits for exploitation, the problem of strategic minerals will only become larger with time.

These factors, combined with the more widespread industrialization of the world economies, will cause the list of strategic minerals to grow. As noted previously, oil was not a strategic material in 1920 for the United States, nor was it until 1970, when the curves of consumption and production finally crossed. But oil is now and will continue, for the indefinite future, to be a strategic mineral for the USA.

In the case of oil, not only for the United States but for other industrialized countries, that it *is* a strategic mineral was strongly emphasized by the military actions in the Persian Gulf, in the late 1980s, as the United States and certain other nations undertook to protect the oil tankers coming from that region. These naval operations cost both money and lives. Avoiding such problems over strategic minerals in the future is one of the challenges that lies ahead for the world.

Finally, the concept of a "strategic mineral stockpile" should be discussed. This mineral stockpile idea stems originally from experiences in World War I and World War II, and commonly it is thought of as a military measure. However, there is the possibility, if not the probability, that major conflict, given the horrors of atomic

211

warfare, will be avoided in the future; and therefore, like building or re-commissioning battleships, we are fighting the last war with the mineral stockpile concept, rather than the next war. As suggested elsewhere in this volume, the next war is really at hand now, in the form of economic warfare, the competition amongst economies to see which ones are the more productive and can do more for their citizens. And, therefore, strategic mineral stockpiles, if such are to be built at all, should be oriented primarily toward civilian economic demands.

If an atomic war should break out, there would be no time available to make use of any strategic mineral stockpile, so, the mineral stockpile, from that point of view, is obsolete. At best, a strategic mineral stockpile for civilian needs could only last a short time. Admittedly, that is what it is for; but in broader terms, it is clearly not the answer and other avenues of keeping strategic minerals steadily flowing through the world economies should be consistently pursued.

By law, the United States is empowered and charged with building a strategic mineral stockpile. This has been done more or less half-heartedly, including the partial filling of an abandoned salt dome mine or two (in the Gulf Coast) with oil. But in the context of today's realities, the whole concept of strategic mineral stockpiles is an anachronism, and the real answer is genuine free trade.

Chapter 17

Nations and Mineral Self-Sufficiency

We have examined the relationships of minerals and nations in a variety of ways, including how minerals have affected the economic development of nations, minerals in warfare, minerals and health, minerals and the money of nations, and how the fortunes derived from minerals, by governments and by individuals, have been spent. But there is frequently expressed what is regarded by many as the most basic question of all: How self-sufficient is a given nation in mineral resources?

This question does not have a simple answer, for it involves several variables. A largely agrarian nation, which has little manufacturing, needs very few minerals directly. So, how self-sufficient a nation is in minerals is a function of its needs, at its particular level of economic and industrial development. Does one view the degree of self-sufficiency in terms of what it might be, if the country were to embark on an industrialization program? To some extent it is a circular situation: A country does not need many minerals because it is not industrialized and it may not be industrialized because it does not have many minerals. But in any event, a nation may be virtually self-sufficient in minerals, at a low standard of living; but if it chooses to rise above that, it may be very short of minerals. A highly-diversified, highly-developed, industrial economy is likely to be short a great variety of minerals. No industrialized nation is mineral self-sufficient.

Thus there is the variable of how much and how fast a nation is becoming industrialized. Also, increased population, with a corresponding expansion in industrial production, may bring about shortages in minerals which were in sufficient supply to take care of the earlier, lesser demand. For example, the United States, through the middle of the twentieth century, had enough oil to supply its needs; but against a rising demand, it subsequently became a permanent oil *importer*. It may be that merely a steady demand for a given mineral will eventually deplete a nation's indigenous supplies, to the point where the mineral has to be imported. Or the high-grade ores may become depleted and better-quality ores from other countries are imported, because they are more economical to use than are the domestic, lower-grade resources.

An example of this, already cited in another context, took place in United States iron ore. Once possessed of abundant, high-grade, 60-percent iron, hematite, the U.S. exhausted its rich ore and the lower-grade taconite, 25 to 30 percent iron, had to be used. There is still lots of taconite, and it is being mined; but foreign, higher-grade ores are now competitive and are being imported. In effect, in the economics of the day the United States lost its self-sufficiency in iron ore. We note here, however, that in some circumstances, at a price beyond world prices, a nation can still remain self-sufficient, or regain self-sufficiency in a mineral. At a price, the United States still is self-sufficient in iron ore. So, self-sufficiency is also a matter of economics.

All these variables, therefore, do not easily lend themselves to very precise statements, as to the degree of national mineral self-sufficiencies. It is a continually changing situation. Therefore, an analysis — mineral-by-mineral and nation-by-nation — becomes a very large task; and once completed, it would be almost immediately out of date. Accordingly, here we will not detail, in any great degree, national mineral self-sufficiencies, but present the situation in broader view.

It is clear from the outset that smaller nations (in terms of both land area and population) cannot very well be considered in this discussion. In terms of population, with a few exceptions (such as Switzerland), if they are small, they will not have an industrial base of any consequence; and therefore mineral self-sufficiency is not a major consideration.

But more importantly, from a geological point of view, one fact emerges quickly. To have a variety of mineral resources, there must be a substantial land area. The geological *occurrences* of various minerals, including energy minerals, one from another, are quite different. Copper and oil occur in quite different geological environments. Therefore, only countries with fairly large areas, which include diverse geological terrains, can hope to have a broad spectrum of mineral resources.

The countries which are seriously concerned with mineral self-sufficiency are the larger industrial countries. This category includes the USSR, the United States, China, Brazil (covering nearly half of South America), Japan, Western Europe (considered as a whole, due to the common market concept) Great Britain, Canada, and perhaps India and Indonesia. South Africa also deserves special note, because of its remarkable array of minerals, and its relatively large industrial complex — which accounts for a sizeable part of the total gross product for all of Africa.

The clear winner in the matter of large and diverse energy mineral and mineral resources is the Soviet Union. In its more than eight million square miles, which makes it the world's largest country, in terms of area, it has the world's largest known gas reserves, the world's largest coal deposits (although it may be that the United States has the largest deposits which can be economically recovered, in the foreseeable future), large oil reserves (about two and one-half times those of the USA), enough undeveloped hydropower to supply the needs of 100 million more people, and a great spectrum of metals. Only in a few minor metals is the Soviet Union lacking. However, the Soviet Union is beginning to face one problem. As the prolific oil fields of the southern USSR are depleted, which is what is happening, it is necessary to move northward, into more hostile territory. As the better oil deposits have already been developed, apparently, the Soviet Union is no longer able to expand its oil production. Currently, demand and supply are close to being equal, at about 12 1/2 million barrels a day. Shortly, demand is likely to exceed supply, so that the Soviets will soon pass from being self-sufficient in oil, to becoming a net importer. In this regard, the USSR is conveniently located not far from the Persian Gulf, site of the world's largest oil deposits.

The very large Soviet coal deposits may have been a major factor in unsettling of the Soviet economy, in the late 1980s. Soviet coal miners went on strike regarding a great variety of wage conditions and benefits. The official news agency, Tass, reported that the strike threatened "catastrophe" in the steel and power industries, very heavily dependent on coal. To satisfy the workers, the Soviet leader, Gorbachev, promised to double the worker's pensions, increase the number of holidays, shorten the work week, and provide a large spectrum of consumer goods previously difficult to find or, in some cases, almost unattainable. These included meat, soap, cloth, shoes, cars, refrigerators, and furniture. But to do this meant taking these already scarce items from other parts of the country, for during the previous year, 1988, shoe production had fallen four percent, and cloth by six percent, meat had been rationed in many parts of the USSR for several years; and for the general population, there was a wait of several years for an automobile.

The settlement of the strike, which threatened to paralyze the country, set a very costly precedent, which tended to further upset the Soviet economy. The ramifications of this situation are still being worked out; but it is clear that the presence of huge coal reserves and the resulting dependence of the Soviet Union on these reserves enabled the miners to pressure the government into actions which it would otherwise not have taken. And thus the Soviet coal deposits rather directly affected the course of the Soviet economy and its political direction. At the time of the strike, the Soviet government already was running (translated into dollar terms) a deficit of about $180 billion, and to meet the miners' demands, the government was forced to simply print money, further exacerbating the already serious inflation problem. Coal was the critical element in this drama and the weapon by which the miners attacked the government.

China already has a fairly large industrial base and is attempting to expand it further. It must try to continue this expansion, because of its still growing population. China does have a wide variety of minerals, including the world's largest deposits of tungsten, a most useful industrial metal, in which almost all other countries are in short supply. China has large iron deposits and the world's third largest coal deposits. It was these latter two resources which impelled the Japanese to invade northern China (Manchuria) in the 1930s. China lacks oil, currently producing only about two and one-half million barrels a day, compared with the USSR production of about 12 1/2 million barrels, and the 7.8 million barrels of the United States. At best, China is expected to produce only about four million barrels a day by the year 2000. Although China now exports a little oil, in order to obtain foreign exchange, the demand for oil in China will exceed supply, very shortly. Indeed, if China were to use oil to replace human labor and use it in vehicles, as is done in the western industrial world, China would right now face a huge oil shortage. For example, there is only about one car for every 2, 000 people in China.

Coal is China's current chief fuel, accounting for about 75 percent of the energy supply; and this situation is likely to continue on to the end of the century at least, and probably much beyond. Though possessed of abundant coal, it somehow seems that China is unable to develop it fast enough to keep up with the demands. Lack of sufficient energy has become almost a way of life, as power shortages idled nearly 20 percent of China's industrial capacity in 1987. The chances of China equaling either the USA or the USSR oil production capabilities appear remote.

China's basic problem, relative to mineral and all other resources, is simply that growth in population has been outstripping the resources which can be found and developed fast enough to meet that demand. Such energy mineral and mineral resources as China may have, will be employed chiefly to keep even with the basic needs of its more than 1.1 billion people. As noted elsewhere in this volume, it has been said that China is self-sufficient in oil; but again that is only in the sense that there is no real attempt to shift the burden of work from the human back to the machine. China, in terms of its population, and against the criteria of western world mineral use standards, is very short of mineral resources, and will remain so.

At the beginning of this century, the United States enjoyed an abundance of virtually all energy mineral and mineral resources needed by a highly industrialized society. And this abundance continued until about the 1960s, when increased demand and lower and lower grade (remaining) deposits of copper and iron, in particular, made the USA increasingly a net importer of metals. Prior to that time, it already was an importer of such metals as cobalt, nickel, chromium, platinum, tungsten, vanadium, aluminum ore (bauxite), and manganese. In 1982, Chile replaced the United States as the world's largest copper producer. And early in the 1970s, the United States became hostage to other nations, in terms of oil supplies.

Brazil is the industrial giant of South America and has tremendous deposits of iron ore, and some good aluminum ore; but both of these take a considerable amount of energy to process and Brazil is very short of energy, particularly coal and oil. This has caused severe problems in terms of balance of payments and has put Brazil in a very large debtor position. During the great rise in oil prices, Brazil had to import almost all of its oil; and the bills still remain in the form of a huge foreign debt. More recently, Brazil has had some minor successes in oil exploration but still does not have nearly enough for its own needs and almost certainly never will.

Western Europe, including Great Britain, enjoyed (in total) a fair degree of mineral self-sufficiency in the early stages of the Industrial Revolution. But it soon became apparent to Germany, for example, that it was not self-sufficient, by any means, to enter the industrial age in a long term sense; and this was cause for its aggressive territorial expansionist plans, which led to two World Wars. Great Britain, although a relatively small country, did have a rather remarkable variety of mineral resources for its size, including coal. All of this being in a fairly limited geographic area worked out well, as the various resources did not have to be transported far. The geology of Great Britain is remarkably diversified, for its less than 90,000 square miles (7,000 square miles smaller than the State of Oregon). Parenthetically, it might be noted that this is perhaps the main reason why the science of geology was first developed in Great Britain, for it is a splendidly compact and natural geological laboratory. However, the demands of the Industrial Revolution soon depleted most of Britain's mineral resources and she turned to her colonies for help, in this regard.

Now, shorn of colonies, Britain is quite dependent on free trade and mineral imports. The relatively late discovery of oil in the British sector of the North Sea has made that country, for the moment, not only self-sufficient in oil but a net exporter. However, this surplus is not likely to persist beyond the mid-1990s, or at the latest, the year 2000.

Canada now has only a modest industrial base, perhaps because it has a relatively small population and is still a frontier area in many ways; but the industrialization of Canada is growing. To support this, Canada has a fairly good variety of minerals, including iron ore in great quantities; and it has huge, undeveloped hydropower resources. Currently, Canada is self-sufficient in oil; but will probably lose that sufficiency within the next decade, although Canada has some of the world's largest oil-sand deposits. But these can only be developed slowly, for the capital costs are large; and the ability to produce oil in quantity from this source (as well as the heavy oil deposits there) takes time to develop and can hardly equal the production that can be obtained from wells. Nonetheless, Canada has energy and mineral resources which should support it well, in this regard, for many years to come. In terms of energy resources, not only does Canada have vast oil-sand and heavy-oil deposits, but a very large hydropower potential, as already noted. Special mention should be made of Canada's uranium deposits, which can only be described as huge. Large uranium mines are already developed in Ontario and Saskatchewan and even larger deposits have been discovered and not yet put into production, but are there for the future. Currently and historically, Canada has been a supplier of raw materials to the United States, in particular, and to other countries. But Canada is growing and will gradually retain and process more and more of its indigenous resources, becoming a significant industrial nation in its own right.

India has a number of large industries, but lacks energy, especially oil, and is unlikely to find major oil deposits in its territories. It has a fairly wide variety of metals, but lacks, as do most other countries, such things as chromium, tungsten, platinum, and cobalt. India and China are very much alike in having very large populations, which (in effect) use virtually all available mineral resources, simply to sustain what is a relatively low standard of living. India, therefore, is "self-sufficient" only in the sense that it does not import a great many minerals, because it cannot afford to, but must live with what it has available. It does have iron and coal, and this supports a fair-sized steel industry. It is quite short of oil, and even with its limited use in India, substantial amounts have to be imported.

218

Indonesia, in terms of population the fifth largest nation in the world, is much like China and India in having modest mineral resources, which are more or less adequate for the demand, but the demand is probably conditioned by the availability of the minerals. That is, if more were available, more would be utilized. Indonesia does have some nickel, copper, aluminum ore, silver, gold, a little iron ore, and tin. Of these, tin ore is the only significant export. The chief mineral resource of Indonesia is oil, which provides more than half the export income for that country. Indonesia is currently the 10th largest oil producer in OPEC and is the largest oil producer in the Far East. However, it does not really have very large oil reserves, with only an estimated 8.4 billion barrels. It is simply that it has more than its southeast Asian neighbors. In terms of the total industrial mineral spectrum, Indonesia is not very well endowed.

South Africa, although it occupies only a small portion of the African continent (some 471,000 square miles, about one-eighth the size of the United States), accounts for more than half of the gross industrial production on that continent. One of the reasons it can do this is because it has a remarkably wide variety of minerals, including such rare items as platinum, of which it holds about 80 percent of the world reserves. Also, South Africa is one of the very few countries in the southern hemisphere that have sizeable coal deposits, which it uses in its steel industry and also to supply about half of its oil needs, by converting coal to oil. This is a fairly expensive process; but South Africa pursues it, for that government wishes to remain mineral self-sufficient as much as possible, fearing economic isolation. Oil is the principal mineral resource which South Africa lacks; and this is not likely to change, except for the prospect of perhaps some modest offshore discoveries. Other than for oil, South Africa is markedly self-sufficient in minerals.

As stated at the beginning of this chapter, we are looking at nations which are large either in terms of territory or in terms of industrial might, or both. Japan has an area smaller than California and a substantial part of it is mountainous and volcanic. In terms of mineral and energy minerals, among industrial nations, Japan is the clear loser. Yet Japan is the second largest industrial nation in the world. It has become this, in spite of the fact that it has virtually no mineral or energy mineral resources. It is nearly 100 percent dependent on imported oil, it has very limited coal deposits, and it has no metals, in any substantial quantity. Japan is markedly deficient in indigenous supplies of both minerals and energy minerals. Japan lives by imports, upgrading them to the finished product, and then exporting them. Without large and continual supplies of raw materials, Japan would be virtually paralyzed, industrially.

The tabulation which follows, from the U.S. Bureau of Mines, presents the degree of dependence of the United States, Japan, the European Economic Community (E.E.C.), and the USSR on the principal industrial metals. 100 percent means total dependence on imports. 0 percent means complete self-sufficiency.

	U.S.A.	Japan	E.E.C.	USSR
Niobium	100%	100%	100%	0%
Manganese	99	97	99	0
Bauxite (aluminum)	97	100	86	30
Tantalum	90	100	100	0
Chromium	88	99	100	0
Platinum	85	98	100	0
Nickel	75	100	100	0
Tin	72	96	92	24
Silver	58	93	93	18
Zinc	53	53	81	0
Tungsten	48	68	100	14
Gold	43	96	99	14
Iron ore	36	99	90	0
Vanadium	14	78	100	0
Copper	7	99	99	0
Lead	0	73	74	0

From this tabulation, and to summarize, it is evident that no nation is entirely self-sufficient in terms of minerals. The USSR comes the closest in this regard; and it is significant that it has the world's largest total fossil energy reserves and is still reporting some major discoveries. One example is the Tengiz oil field, first drilled in 1979, and still undergoing exploration, with excellent yearly results. By 1988, oil reserves found to that date were estimated to be at least 18 billion barrels — about twice the size of the Prudhoe Bay field in Alaska. With the aid of this and its many other large fields, the USSR is now the world's largest oil producer at 12 1/2 million barrels a day. However, they are probably unable to increase this very much, as discoveries are being offset by declining production in the older fields. They may, however, be able to hold this peak for a time. Soviet gas reserves are still growing and they are even now very large, being at least the energy equivalent of 200 billion barrels of oil, which quantity is more than three times their oil reserves.

However, the Soviets still lack a few minerals, silver and bauxite (aluminum) being two of them and both critical in defense industries. Nevertheless, in terms of over-all mineral supplies, the USSR is exceedingly well endowed, and has the advantage of still having a large amount of territory yet to be explored in detail. Siberia is a difficult terrain. Keep in mind that the United States, including Alaska, has an area of about three and a half million square miles, whereas the USSR is eight million square miles. The more territory, the more chances for finding minerals.

Much of this great Soviet land mass is part of Siberia, and Siberia is a difficult terrain in which to search for and develop minerals. Access has been poor, but the Soviets are building roads. They also make extensive use of helicopters to transport their staff of exploration geologists, which is several times larger than that of the United States; and the Russian geologists are making discoveries. To augment this effort, the recently completed Baykal-Amur-Magistral railroad (the BAM) has opened up a large new territory, from which minerals can now be shipped. That the Soviet Union has made excellent progress in its avowed program to be mineral and energy mineral self-sufficient, is attested to by the fact that, in 1928, the USSR was 60 percent dependent on imports for its mineral needs. Currently, it is almost entirely independent of outside supplies.

Viewing the world at a whole, no mineral supply shortages are likely to occur in the near future, given a free trade situation (see next chapter). And if ultimately large quantities of cheap energy could be made available (perhaps from fusion), the mineral supply problem would be greatly helped, for with cheap energy the large, low-grade mineral deposits which cannot now be mined economically, could be recovered. However, the other industrial nations, beyond the Soviet Union, are now substantially — and will be increasingly — dependent on off-shore sources of raw materials. Where minerals have been placed by the geological processes of the past, and where these minerals are now in demand, are, in many cases, quite different places. In the next chapter, we consider whether free trade can overcome the irregular mineral distribution problems which geology has caused.

Chapter 18

International Access to Minerals — Free Trade Versus the Whims of Geology

We have seen that mineral resources are geographically distributed quite unevenly on the Earth. When this reality encounters the boundaries of nations, the result frequently is a substantial difference between the places where the resource is used and the source site of the mineral. This causes problems. In the past, the possession of certain minerals at times has meant (essentially) the difference between life and death — for example, copper and tin, to make the bronze weapons which the Romans used so effectively on their less well-armed adversaries (which is why the Romans wanted control of the tin mines of Cornwall).

In our present society, the possession of — or access to — minerals, especially energy minerals, largely determines our material standard of living. In the past, access to minerals was frequently settled by warfare; and, in fact, this has even occurred quite recently. The matter of oil supply, as we have already discussed, was the immediate cause of Japan going to war in 1941; and Japan's general expansionist program in the 1930s was primarily to secure access to mineral resources, especially the coal and iron ore of Manchuria.

At the present time, the unequal distribution of minerals is being accommodated by peaceable trade. Will this always be true in the future, or will the history of conflicts over mineral resources be repeated? When the Arabs and Iranians shut off oil supplies to the United States in the 1970s, the cry was heard, "send in the Marines!" Violence occurred in the gasoline lines that the shortage created in our supposedly civilized United States.

If free trade can be maintained and if crucial economic problems are resolved (chiefly a matter of "balance of payments," and "values of currencies"), perhaps the erratic distribution of minerals can be compensated for without warfare. However, from time to time there have been attempts to disrupt free trade by means of cartels — a cartel being defined by Webster as "a combination of independent commercial enterprises designed to limit competition." It was an Arab-Iranian cartel which precipitated the thought by some of taking military action in the oil crisis of the 1970s. Reviewing the history of some cartels is worthwhile, to see if they can, for a protracted time, limit trade.

There are two kinds of mineral cartels: One is natural and one is artificial. A natural cartel is one that is the result of geological processes which have created a relatively limited distribution of a particular resource. An example of this would be platinum, where South Africa is particularly well endowed with this critical metal; and the rest of the countries have very little or none. The ability of a single nation, or perhaps two or three that have such a natural cartel or monopoly to control prices, is considerable. It may be challenged only by the ability of the consuming nations to discover or develop effective substitutes.

An artificial cartel is one in which the resource is reasonably widespread but where the countries holding a substantial portion of the resource try — artificially — either to hold back production in order to bolster the price, or simply and collectively raise the price. The most famous (or infamous, depending on your point of view) example of this recently, has, of course, been OPEC — the Organization of Petroleum Exporting Countries. Another example, not so well known, has been the tin producers cartel, which, however, came apart in 1985.

Natural cartels may well succeed, but artificial cartels almost always seem to find themselves in disarray, at some time or another. The pattern of cartel rise and fall seems to be a fairly standard one. First, the cartel will push prices up beyond what the normal price would be. This encourages more production by marginal producers and smaller suppliers, outside the cartel. It also encourages substitution and conservation by the end users. The cartel will then generally try to hold down supply, by agreeing among themselves on individual quotas. This almost invariably does not work. Because of domestic political and economic demands for money, from the sale of the particular commodity, individual members of the cartel will tend to cheat; and ultimately the whole cartel is in disarray. This is what happened to OPEC, in the drop of oil prices during the mid-1980s.

First, the steep rise in oil prices stimulated development of new (and previously marginal) prospects outside of OPEC. These came particularly from the United Kingdom and Norway (North Sea area), Mexico, and the Soviet Union. These countries had never joined OPEC but they soon became significant exporters of crude oil, to compete against the oil cartel. Second, additional natural gas production became economical with the increased price of oil, and gas began to displace the markets previously held by oil. The use of coal and nuclear energy (to generate electricity) further reduced the demand for oil in many countries, including the United States. Third, rising prices created a surge in oil exploration in the United

223

States, in particular. The number of wells drilled each year rose annually for eight consecutive years starting in 1974 with 33,000 wells, and peaking in 1982 at 88,000. Fourth, rising prices encouraged conservation efforts throughout the oil importing world. Large sums of money were invested in factories, homes, and other buildings to install devices that conserved fuel. And a new generation of much more fuel-efficient cars began to replace the old gas-guzzlers. There was also a recession in the early 1980s which reduced demand for oil. Demand for OPEC oil dropped 12 million barrels a day, between 1979 and 1982.

Finally, as demand declined, OPEC was forced to reduce production, in order to maintain oil prices; but some of the poorer countries, with large populations, such as Nigeria, decided to use their idle oil production capacity and increase their oil income; and they began to cheat on the cartel's official price, and offered discounts and produced oil beyond their quota.

For awhile, Saudi Arabia, with the largest potential oil production, single-handedly tried to support oil prices by cutting back on its own production, which in 1980 was about 10 million barrels a day. They had cut it back to about 2.3 million barrels a day, by the summer of 1985. But the Saudis warned the other OPEC members that they would not indefinitely be the "swing producers," in order to maintain prices. In the fall of 1985, the Saudis began to run out of patience and also money. Ultimately, the cartel began to come apart, as the Saudis increased production and OPEC members abandoned efforts to hold up prices. Each went out for a "fair share" of the oil market, regardless of price. The price of oil came down rapidly, from about $28 a barrel to less than $10, briefly, in early 1986. The price subsequently seesawed back and forth, gradually working higher to a more stable price of between $18 and $20 a barrel by 1989. But, basically, there was a world oil glut for the moment.

Although the short term effect of the price cutting by Saudi Arabia was devastating to the income of the oil-producing nations, the architect of this program, Sheik Yamani, the Saudi oil minister, did strike a fairly effective blow toward ultimately putting OPEC back in the driver's seat (some time in the future) on oil prices. They did this by cutting the price and by keeping up production in the face of shrinking demand so that the marginal wells in many parts of the world, particularly the United States, could not be economically produced. About two million barrels a day (at that time) of the approximately nine million barrels of daily U.S. oil production (which does not count about a million and a half barrels of condensate liquids from natural gas each day), were from "stripper" wells, defined as wells which do not produce more than 10 barrels a day. The reservoirs, mostly sands, from which these wells produce are relatively tight and have low pressure. The result is that when the pumping of these wells is stopped, because they are no longer economical, due to the low oil price, generally they cannot again be put back into production. They are lost permanently.

During the interval of low oil prices in the mid- to late 1980s, a considerable amount of stripper-well production was lost, putting the United States in a less competitive position, once the demand for oil increased. In this way, Sheik Yamani and OPEC did make a long-term, negative impact on U.S. oil production, and put OPEC in a stronger position, with respect to the oil producing ability of the United States. This was part of Sheik Yamani's strategy and it was a good plan. But perhaps because it was a longer term strategy and the immediate effect of lower prices was to greatly reduce OPEC's income, Yamani came into disfavor with his policy and ultimately was fired from his job in 1986. However, some time in the future OPEC will benefit from the oil production that was lost in other countries, due to the low oil prices of Yamani's administration.

As to the future, an examination of OPEC and the other nations which may compete with OPEC is instructive. OPEC is the classic cartel attempt of the twentieth century, and the final word on its success or failure is not yet in. Sometime in the future, it is quite likely to work better than it does at present, because more and more world oil production will ultimately be concentrated in the Persian Gulf area.

OPEC membership grew gradually from its beginning in Venezuela, in 1960. At that time Venezuela, Iran, Iraq, Kuwait, and Saudi Arabia established OPEC. Qatar joined in 1961, followed by Indonesia and Libya in 1962, the United Arab Emirates in 1967, Nigeria in 1971, Ecuador in 1973, and Gabon in 1975. Currently countries which export more than 100,000 barrels of oil per day and which are non-OPEC countries, and therefore OPEC's Achilles heel, are Canada, Mexico, the United Kingdom, Norway, Angola, Egypt, Oman, the USSR, China, Malaysia, and Brunei.

In looking at current crude oil production and the potential for increasing oil production, however, it is clear that the non-OPEC countries are in a weaker position than are the OPEC nations. Of the non-OPEC countries, Mexico, Angola, Oman, China, Malaysia, and Brunei might be able to increase production from current levels, but only slightly. In the case of the two biggest non-OPEC producers by far, the United States peaked in production in 1970 and has declined considerably since then; and the USSR is now in the process of peaking. Neither country can appreciably increase its oil production.

In contrast, Saudi Arabia currently is producing less than five million barrels a day, but has a potential of 10 million barrels a day (or more), by just opening the valves. It also has more areas to explore, including deeper, undrilled zones in existing fields. Thus, Saudi Arabia by itself, in the OPEC group, has the ability to strongly influence, if not control, oil prices for years to come and provide the member nations of OPEC with an optimistic future.

Kuwait and, to a lesser extent, Indonesia, also have potentials for increasing oil production by which to influence prices. And as things settle down between Iraq and Iran, production in those two nations could substantially increase, giving them leverage on world oil prices, along with their fellow OPEC members. However, to repair the war damage, both Iran and Iraq are likely to over-produce (relative to their OPEC quotas) for a time, more or less ignoring OPEC pricing and quotas.

The OPEC cartel will probably limp along until the more marginal non-OPEC producers have depleted their resources. At that time, the cartel will evolve from being an *artificial* cartel, which is chronically insupportable (as has been shown currently and in the recent past), to a *natural* cartel, which has a much better basis for survival. With more than 60 percent of the world's oil in those countries bordering the Persian Gulf, all of whom belong to OPEC, a natural cartel is going to eventually arise. This resurrection of OPEC will be a gradual process but a certain one. And when it does become a *natural* cartel, it will be much more powerful than previously.

Other cartels, however, are not so likely to meet with such ultimate success, because the resources involved are not used so widely and in such volume, and are amenable to substitution more readily than is oil; moreover, the resources are not so concentrated geographically. The International Tin Cartel, in the late 1960s and early 1970s, raised the price of tin very substantially, with the result that non-members, such as Brazil and China stepped up production. Even Great Britain was able to open up some of the long-closed, ancient tin mines of Cornwall, at one time worked by the Romans. And tin users found they could very well turn to substitutes, including plastics, glass, cardboard, and aluminum. In 1972, about 80 percent of the beverage cans in the United States were made of tin plate. However, by 1985, virtually all American beer cans and 87 percent of the soft drink cans were made of aluminum. Tin lost out.

The cartel idea goes back at least 3,000 years; but in almost every case, the free market prevailed. The wheat, uranium, and sugar cartels are all history. The rubber cartel bounces along and is not doing very well at present. Coffee cartels have been tried but seem less effective in controlling prices than the temperature in Brazil. A hard freeze does more than price agreements to bolster the market. Conversely, good weather usually creates a crop surplus, which cannot be stored indefinitely and, when sold, depresses the price.

There is, however, one cartel which has been reasonably effective. Even though a zircon or the synthetic gemstone, cubic zirconium, both of which quite closely simulate the appearance of a diamond, can be bought for a fraction of the cost of a genuine diamond of the same size, the concept that "diamonds are a girl's best friend" seems to have prevailed. This view is heartily endorsed both by young ladies and De Beers Consolidated Mines. The diamond cartel is essentially one company — De Beers — because they control (and have controlled for more than 50 years) about 80 percent of the World's diamond production.

Two other minor sources of diamonds, Zaire, and the Soviet Union, have made half-hearted attempts to break the cartel; but they are such small producers that the effect of their efforts has been negligible and seeing that, for the most part, these other two sources have simply fallen in line. What makes De Beers successful is the fact that it is one of the best *natural* cartels, for gem-quality diamonds are found in only a very few places in quantity and De Beers controls the majority of these sources. Also, De Beers is virtually the only company in the business. If there were two or three major diamond producers, this might make a difference, as total agreement — at all times — would be unlikely; but as there is only De Beers, it has only to agree with itself. It is not subject to differences of opinions and diverse economic circumstances, as in the case of OPEC, from time to time. So geology and young ladies (and older, rich ladies, too) have combined to help insure the success of this cartel. Any efforts by the young men of the world to break this cartel, so far, at least, have not been very successful.

There are certain other minor, strategic minerals, such as vanadium and cobalt, which are now in natural cartels. But none of these is a major item of commerce; and to a considerable extent, the nations holding these minerals are in rather urgent need of foreign exchange, and therefore not likely to cut off shipments for very long.

Looking at copper, aluminum, iron, and other more extensively-used metals, they are so widely distributed that it is very unlikely an effective cartel could be formed. There was, in fact, a brief attempt, in the 1970s, to form a bauxite (aluminum ore) cartel. This cartel was short-lived, in large part because of the huge deposits of bauxite which had been discovered in Australia. This catapulted Australia, which had hardly figured in the bauxite trade previously, to being the world's largest producer. Suriname and Jamaica lost heavily in this situation. And, again, mineral resources and who had them, made a critical difference to some nations.

It seems probable, therefore, that free trade in minerals can be a fact of international life for some time to come, if political/economic barriers are not erected. However, the pressures for such barriers will be present, for the combination of lower wages in countries which have higher quality and more abundant mineral resources tends to cause loss of jobs in the higher-cost countries. In 1982, the United States lost out to Chile, as the number one world copper producer. Lower Chilean wages and their higher grade ore caused this, and many United States copper miners lost their jobs permanently. Similar situations will continue to occur in other minerals, and among other countries.

In terms of cartels, only one major resource cartel is likely to be viable, if the members wish it to be. As we have noted, that one is OPEC; but actually only some of the present members of OPEC will be in the final OPEC cartel — those countries which are in the Persian Gulf region. The other members have substantially less oil reserves and will not be able to stay in the oil game nearly as long as will the Persian Gulf players. The Persian Gulf portion of OPEC is projected to become a strong natural cartel, beginning about the year 2000 or 2005, as non-OPEC members' production dwindles. However, international political and economic circumstances at that time may make the cartel only moderately effective. The world is becoming so economically interdependent that it is doubtful whether the OPEC countries around the Persian Gulf would enforce their cartel to a disastrous extent, because they already have large investments in the countries where they must sell their oil. Substantial injury to the economies of these countries would not be to OPEC's interest. Furthermore, looking very long term, when the final portions of OPEC run out of oil — which means nations in the Persian Gulf — they may have to depend quite substantially on investments made in other countries, with more diverse economies.

228

In summary, free trade can erase, or at least ease, the whims of geology, in regard to placement of mineral and energy mineral resources. But free trade will also adjust the balance in standards of living (at least to some extent) among countries. The resource producers' standards will probably rise with a concordant lowering of the standards of living in the consuming countries. The great transfer of wealth which has taken place recently and will probably continue to take place, from the industrialized nations to the major oil-producing countries, is an example. But it has become rather clear that going to war over access to mineral resources has not been an ultimately satisfactory solution to the problem. In some fashion, free trade must prevail in our world of the future.

A final note should be made, however, that this discussion has been related to free trade in minerals and energy minerals. This is a case of where the material involved is possessed by some countries and is needed by other countries. The concept of total free trade for all products is another matter, textiles being an example; these being products which can be produced by many countries, among which there are wide differences in wages and standards of living. If total free trade were instituted, the resulting loss of jobs and reduction in the standard of living in some countries would create a political outcry that would, and, indeed, has to the present time, precluded a totally free trade policy. But in the case of vitally needed energy minerals and minerals, free trade must be allowed to prevail, to at least some reasonable extent. Still, using the leverage gained from the uneven distribution of mineral and energy mineral resources, as a political weapon, as was done by the Arabs with their oil (against the United States) is always a possibility. And as mineral resources become more important and more scarce, there will no doubt be a temptation to do this at times — which, in turn, could create dangerous crises situations. Hopefully, however, holding mineral resources hostage, as a political weapon, can be avoided in the future. This is another challenge in the realm of minerals but, more importantly, in the realm of basic human relationships. We must learn to live together on this finite globe. Perhaps the varied geography and geology of energy minerals and other minerals will help us to do this. We need each other.

Chapter 19

Minerals and the Future

Everything we have, including ourselves, comes from the Earth. All tools, machines, and the most of the energy to run these machines — to produce the necessities, as well as the luxuries of modern society — ultimately come from the rocks and soil beneath our feet.

The human race inherited a storehouse of materials produced in the Earth by myriad geological events, over great spans of time. It is an inheritance which nature took billions of years to accumulate for us. For hundreds of thousands of years, the human race slowly evolved and drew very little from this treasure storehouse. Generations came and went and virtually no mineral or energy mineral resources were used. The Earth was left much as each generation found it. But then the arrival of the Industrial Revolution, combined with advances in education, technology, and medicine did two things: The growth in population greatly accelerated; and the use of mineral and energy mineral resources rose exponentially, and this continues. More and more people are using more and more minerals, including energy minerals, at an increasing rate. Far more energy and mineral resources have been used in the world since 1900, than during all previous time. In the case of oil, the first 200 billion barrels of oil in the world were consumed between 1859 and 1968; but it took only the following 10 years to consume the second 200 billion barrels.

This phenomenal rise in the production and consumption of energy mineral and mineral resources is an event unmatched in all history, and where it will lead no one can really tell. It does seem clear, however, that among the various resources we are considering, energy is the most important: For without energy, other mineral resources cannot easily be obtained, nor effectively used, once they have been obtained. It has been said that energy is the key that unlocks all other natural resources. To find a piece of native copper and, by means of a stone, beat it into a crude knife, as was done by many early humans, is useful only to a small degree. To mine and process millions of tons of copper ore into thousands of miles of electric wire, and many kinds of electrical equipment; and then energize these wires and equipment with power produced from copper-wound electric generators, makes copper an infinitely more useful material than when it is simply hammered into a knife. Energy sources are far more important to world economies than their deceptively simple percentage of the gross national product. Without energy, the rest of the economy would hardly function at all.

So what about energy resources for the future? The petroleum interval will be brief. The United States has become a net importer of this resource and will continue to be, for the indefinite future. The USSR is currently peaking out in oil production. More and more, the production of oil will be concentrated in the Middle East, a good reason for the aggressive pursuit (by the major industrial nations), of alternative energy supplies. This will not be a simple task, for, in general, the economics of alternative energy sources are not nearly so attractive as those which, historically, have involved merely drilling for oil. However, even drilling for oil and mining for coal increasingly is showing a poorer ratio of energy-recovered to energy-used in the recovery.

Currently in the United States, about 25 percent of energy produced is used to produce *other* energy; that is, to drill for oil, mine coal, mine uranium, cut wood, make solar energy conversion devices, and so on. As the easily found surface and near surface energy mineral resources (petroleum, coal, uranium) are produced, the cost in energy to produce more energy is estimated to rise to about 33 percent by the year 2000, and may continue to rise after that. This is a trend in energy costs which must be arrested, for if it ultimately *costs* as much in energy as the energy *produced* by the effort, we have no energy profit to use for other energy consuming sectors of the economy.

The world now has a varied energy mix; but initially the energy mix used was hardly a mix at all, being chiefly wood, and other biomass, such as grass, twigs, leaves, and dung. But gradually, coal, oil, gas, hydropower, wind, solar, geothermal, and nuclear power have been drawn upon, although not in the same ratios in the various countries. For example, in the industrialized nations, petroleum is the major energy source; but in some countries, wood continues to be the most important fuel. France gets 70 percent of its electricity from nuclear plants, whereas New Zealand uses no atomic power but, instead, chiefly relies on hydropower and geothermal energy to generate its electricity.

Currently, in the United States, the energy mix is: oil 44 percent, coal 22 percent, natural gas 21 percent, nuclear power 5 percent, hydroelectric power 4 percent, and other (wind, solar, geothermal, wood) 4 percent. In 1984, nuclear power production surpassed hydroelectric power in the United States. Hydroelectric power production will probably continue to decline (percentage wise) as U.S. energy demand grows, for there are almost no major hydroelectric sites remaining to be developed. Fifty years from now, or perhaps in just twenty years, the energy mix will be different for the United States, from what it is today. And the energy mix for the United States and all countries except, perhaps, the few remaining which still use chiefly wood, will continue to change.

This on-going evolution of energy mix presents a number of challenges for the future. Can one energy source effectively replace another, both in terms of volumes involved and in convenience for end use? For example, can electric power really replace gasoline for our cars, on a large scale? Where will these substitute energy sources come from; and if they are not produced domestically, how can a nation pay for the imported energy it needs? Can foreign sources be relied upon? Will greater or lesser changes in life-styles occur, and if so, what sort of restructuring of industries and societies will take place?

One of the more important things to consider is the effect upon a nation as it *loses* its energy mineral and mineral bases. The American public is becoming only dimly aware of the fact that as more and more minerals have to be imported (oil being the best example), industries and jobs, and the jobs peripheral to those industries, are all lost. This results not only in an increasing balance of payments problem, but, concurrently, an erosion in the value of currency, if exports cannot ultimately balance this drain on a nation's monetary position. One cannot print money indefinitely, in order to pay for material things. Sooner or later the suppliers will resist. And with it all, goes a general loss of economic might and world prestige. It is a slow process, and the citizenry may fail to notice what is happening for a time; but eventually, the standard of living must suffer, which will be felt by all. And thus the destiny of the nation is affected.

The past hundred years have seen great changes in lifestyle. The widespread use of oil, chiefly to power cars and trucks, has been responsible for a substantial part of this change. The daily, massive migration of vehicles into and out of city centers such as San Francisco, Seattle, New York, Houston, Los Angeles, and Chicago, for example, is a phenomenon (in some places approaching a nightmare) which could never have been visualized the day Col. Drake struck oil. But the present scene in this regard is changing, as we are already seeing the one and one-half ton automobile in the United States gradually being replaced by a more modest car, chiefly imported from Japan, where they learned energy economy early. And mass transit, too, is making a re-appearance.

Buildings will be remodeled or entirely replaced with more energy-efficient structures, for our buildings — offices, homes, factories — and not our transportation system, are the largest single consumer of energy in the United States — about 40 percent of the total. The savings in rebuilding the United States (and many other nations of the world) to more energy efficient standards will be considerable, and, indeed, it will have to be done. In myriad ways, energy costs and the major kinds of energy, in the energy mixes of the future, will affect the basic living habits and economies of nations. Except for water, energy minerals are the most critical world mineral resource. An increasing amount of technology, money, and political attention will surely have to be directed toward this vital concern.

One thing about energy is certain: Almost all minor alternative energy sources, and also the principal energy materials currently used — coal, oil, and natural gas — will cost more and, in some cases, much more in the future than they do today. Energy will take an increasing percentage of national and personal budgets. Success of fusion technology may have the potential to change this outlook, but that is unlikely for many decades to come.

We have been emphasizing the importance of energy for those who use it. Turning to the producers of energy, it is already set in stone — the rock reservoirs which hold the Earth's oil and gas resources — that the Persian Gulf countries increasingly will be the focus of world attention, whether there is a war going on in that region or not. There will be milder or more severe forms of economic warfare going on in that region, for as long as there still remain substantial supplies of oil and gas there — which probably means for the next hundred years or perhaps more. Political maneuvering and economic problems will continue to arise in and from this region, as the oil pours out and the money pours in. The challenge will be to keep things on this level and not escalate them to military action. For the next several decades at least, the Persian Gulf nations will be increasingly prominent in world affairs. The *transfer of wealth* from the industrialized nations to the oil-producers will continue, and will have profound effects on both parties.

But beyond those times, there will be written another chapter in history, the chapter which tells of the changes in the nations of the world, as oil supplies and oil revenues decline and finally become virtually non-existent. Although the human race at times believes itself to be beyond the old laws of nature, such will still ultimately prevail. And the story of petroleum is quite likely to be the same story of resource discovery, abundance, use, and decline, which has been repeated many times in the past. An abundance of a particular life-sustaining resource is discovered. The consumers flourish and multiply and multiply again. Eventually, however, the region is depleted of its resource and the consumers must contract their numbers.

The situation is similar to the abundant use of fertilizer on a garden. The plants grow and multiply and flourish as long as the fertilizer is applied; but when it is no longer available, the garden goes back to its normal basic ability to support the plants. The challenge before the Persian Gulf and other currently oil-rich nations (such as Brunei) is to put in place economies, technologies, and investments which will survive the departure of oil, even as great changes were made in those countries with the coming of oil.

But the eventual departure of oil also is a challenge for the ultimate consumers to put in place new technologies and social organizations that will adapt to the alternative energy sources which will have to be used.

There is no parallel in history for such a rapid development and use of a resource, as in the case of oil — and the profound changes which oil has caused. But it is well to keep in mind that the changes will work both ways, with the *coming* of oil — and the *going* of oil.

What has just been said about oil also applies to minerals in the sense that a number of nations are *one-resource countries*, in terms of what they can export to obtain foreign exchange. What happens when this source of money to buy the things not produced locally dries up? Smaller nations face this problem acutely, for, in most cases, all manufactured goods, especially specialized products, such as medical equipment and sophisticated electronic gear, must be imported. This will be a matter which will have to be continually addressed by the wealthier nations on a *humanitarian* basis, rather than strictly economic terms.

In the case of both energy resources and hard minerals, as these resources are depleted in one area, production will shift to other regions; and so there will not only be a change in energy mixes and mineral mixes (as substitutes are introduced) in the consuming nations, but there will also be changes in sources of supplies. And it should be noted again that as the economies of the raw material countries mature, they will more and more tend to upgrade their raw materials at home and this will impact the refining industries in the importing countries.

We have cited the growth of petrochemical industries in the nations of the Persian Gulf region and the development of aluminum plants in Venezuela, close to the source of the ore. To this we might add that for many years, even though the United States did not have much bauxite itself, this ore of aluminum was shipped to the United States for processing and the USA was an exporter of the metal. But, increasingly, the refining of aluminum ore is done by the countries which have the big deposits, such as Australia. Now the United States is a net *importer* of that metal, and some aluminum plants in the Pacific Northwest of the United States lie idle. Another example is Zimbabwe's development of gold refining capabilities in 1988. Previously that country had to ship out gold ore concentrate for refining, which was done by Western Development Corporation, at Perth Mint, in Australia.

Thus, the *raw materials nations* gradually take away industries and jobs from the *consuming nations*. And it will continue to be an ongoing series of adjustments for both parties, with the nations which *have* the minerals tending to gain economically at the expense of the *mineral importing nations*.

There is another fact to be considered. The use of energy minerals and minerals has been markedly concentrated in the past, and to the present time, in a relatively few nations, representing a relatively small fraction of the world's population. At the present time, world oil production is about 57 million barrels a day. The United States, with six percent of the world's population, uses about 30 percent of the world total. Can this continue?

In the industrial world, the chief metals are aluminum, chrome, cobalt, copper, iron, lead, manganese, molybdenum, nickel, platinum, tin, tungsten, vanadium, and zinc. Just since 1950, the use of these materials has increased between 100 and 500 percent. With somewhat less than 30 percent of the world population, the industrialized countries used more than 80 percent of these metals. It is unlikely that these percentages will be maintained, either by the United States, in terms of oil consumption, or by the industrialized countries, in terms of metals. The trend of the raw material producing countries to keep more and more of their product at home, to upgrade it before shipment, tends also to be accompanied by an increased use of that product at home, oil products being a particularly good example. Combining this trend with the more rapidly growing populations in these countries, as compared with the industrialized nations, the conclusion is that there will be an increased diffusion of industrial development and use of energy and energy mineral resources in the future.

For decades, the raw material producers were content to be just that. But that changed after World War II. First, it was the British Colonial Empire that came apart, as the international sport became twisting the British lion's tail. This was, in part, concurrently and then immediately followed by the game of plucking the tail feathers out of the American eagle, as the international raw-materials-producing, economic empire of the United States came under attack.

Kuwait took over Gulf Oil properties. "American" was rubbed out of the desert sands of the "Arabian American Oil Company." Iran, with the departure of the Shah, took over all foreign oil interests. Peru nationalized (without compensation) the company I worked for, International Petroleum Company, Ltd., an affiliate of what is now called Exxon Corporation. The tin mines of Bolivia were nationalized. Many of the South American copper mines were taken over by the governments. Creole Petroleum, in Venezuela, also an Exxon affiliate, was nationalized. In Africa, the nationalization of foreign mineral operations also proceeded.

This is not to unduly criticize these events. They were an inevitable outgrowth of colonialism and reflected a rising sense of national identity in these countries — tempered by a growing realization of what their mineral resources could be worth to them. Some of the actions were quite arbitrary and, in a few cases, reflected outright theft, but it was inevitable. The new government owners of these operations may or may not be as efficient — or even as fair — as were the previous owners; but the pride of the country was on the line, and their response enabled them to say "our minerals are now ours." There will probably be a continuing trend toward nationalization of what remaining foreign interests exist, or perhaps "joint operations" where capital and technical expertise are needed — *for awhile*.

On the assumption that there will be no great rush toward industrialization by third world nations, but that the status quo will be gradually modified in that regard, there will probably be no immediate or great shortages of minerals or energy minerals. However, they will become relatively more expensive as the richer and more accessible deposits are used. But what *is* changing is that the industrialized nations of today are having to play on a different economic and political field than in the past. This field has been created by a rising tide of nationalism and more nearly reflects the geographic distribution of energy minerals and minerals — another aspect of how mineral resources (and where geology has placed them) affect the destinies of nations.

The only thing constant in all of this is change. The human race is an extension of nature's long arm and the great lesson to be learned from the fossil record is that those organisms which adjust to change, survive, and those that do not, become extinct. Adaptation and survival will be the story of the future, in terms of energy and mineral production and consumption. As populations increase against resources which, in many cases, are declining in both quantity and quality, the problems of adaptation to changing circumstances will increase much more rapidly than in the past. Again, it must be emphasized that the demands of a large and rising population, as compared with the past, are so very great, in terms of resource consumption, that in a few decades we will be faced with mega-events unknown previously. We have gone through an era of unprecedented abundance and quality of energy mineral and mineral resources, which we inherited from the literally billions of years of geological events that it took to produce these resources. We are going through them in a geological instant. We have used — and are using — the best, in terms of both quality and ease of recovery. Resource by resource, the trend is turning to a less affluent situation.

In all the discussions we have had concerning mineral and energy mineral resources and how they affect the present and future of nations and individuals, one point has been touched upon only lightly but it deserves greater emphasis. It is the matter of population. Even now this problem is visible in many contexts. It is expressed in the pollution of rivers, lakes, oceans, and the atmosphere. It is visible in the greatly increased rates of extinction for organisms, as human habitation has displaced wildlife habitat. Even with the best of intentions, such as the setting aside of game preserves, as, for example, in Africa, when people are hungry or economically hard pressed, these sanctuaries are invaded and wildlife destroyed.

Just to provide a small amount of a simple energy source, such as wood, forests are being cut down in South America, Asia, and Africa for firewood. More and more destructive floods are beginning to occur, as a result, in many areas. Deforestation of the land is causing severe erosion; and, in many cases, the damage is irreparable in the foreseeable future.

In terms of energy and other mineral resources, the problem is to provide the world's people with a dependable supply of these basic commodities on a scale enabled them to live in some degree of comfort. Ultimately, this has to be accomplished through an essentially renewable, recyclable, steady-state economy. Accomplishing this necessitates that the population be stabilized. It cannot be done when the population continues to be a moving target, in an upward direction. It has been truly said that "whatever your cause is, it is lost without population control." Without population control, there will sooner or later be an inevitable collision between the human race and what nature can provide. It will come in many ways and many places, but it will come. In some areas, this seems to have occurred already, particularly in Africa, India, and China.

Although some countries are making strong efforts and are doing reasonably well to control population growth, the world, in total, does not seem likely to achieve zero population growth in the near future. More recently, China, which had been vigorously trying to control the growth of its 1.1 billion population, has had to admit partial defeat on the project — not a good omen for the future. The dire fact is that world population is growing at the rate of 80 million people a year — a country the size of Mexico, in effect, is being added each year to the world; and there seems no early end in sight to this trend. Although the view is sometimes heard that the world population is beginning to level off, this is only true in some industrial countries; but it is not the situation for the world as a whole. The World Bank predicts that the current world population of slightly over five billion will rise to 10 billion by the year 2050.

The effect of an increasing population is that the demand for energy minerals and other minerals will continue to grow. There will be a corresponding and increasing worldwide concern about supplies of these essentials. We will become much more occupied with obtaining basic materials for survival in the future, than we have been in the current and recent affluent past. We will, and indeed we are now, seeing a world of plenty becoming a land with more modest and more expensive mineral and energy resources.

One of the pronounced effects of this increased use of mineral resources and the fact that the various components of the industrialized nations' mineral supplies must come from widely diverse sources is to emphasize the increasing inter-dependence of all nations. By the same token, nations must now be concerned with global problems beyond what they have been in the past. What happens in one nation, several thousand miles away, may be of vital interest to a country. The concern shown by nations like Japan, Italy, France, and the United States, with regard to events in the Persian Gulf region during the 1980s — i.e., the war between Iraq and Iran, and the free flow of oil — is an example. Thus, mineral inter-dependence may make us "one world" beyond what any political statements can accomplish.

238

Balancing population against available mineral and energy resources must be the ultimate objective. We must develop renewable and reliable energy sources, in quantity. This, together with technology for efficiently recycling metals — with only a modest amount of new material annually drawn upon from the Earth — will, we can hope, lead to a sustainable balance between population and mineral and energy mineral resource supplies.

Given the global mineral interdependence of nations, the matter of how large a world population could be sustained in reasonable comfort, by means of a renewable, recyclable natural-resource economy, with only a small permanent withdrawal of energy mineral and mineral resources, should now be of universal study and concern. And, in fact, it has been given some study. With the presently available resources and technology as a basis, the figures differ considerably; but most of them arrive at the rather sobering conclusion that the population size would have to be somewhat less than at present.

Clearly the challenges of population and resource balance, which lie ahead, are very large. The human race would do well to put less of its energies and resources into strife and division, and more into the common goal of a reasonably affluent life for the populations all over the world that now exist and for those which are to come.

Chapter 20

The Ultimate Resource — Can It Secure the Future?

We entered the Soviet Union through the Port of Nakhodka, east of Vladivostok, as Vladivostok is a major naval base and not open to foreigners. We travelled the length of Siberia and across European USSR, all the way to Moscow and Leningrad, and then back to Nakhodka. The general living conditions and the state of economic development that we saw can only be described as substandard, by western criteria. Yet the Soviet Union has more mineral resources, in total, than any other country. It has the world's largest coal deposits, the world's largest known natural gas reserves; and it is currently the world's largest oil producer. A few miles south of Novosibirsk, the "Chicago" of Siberia, we were taken to see the great Siberian research center; and, as a geologist, I was interested in the Minerals Building. Here we were told about the mineral wealth of the Siberian Shield, matched only, perhaps, by South Africa.

A month later we were back in Japan. Japan has virtually no oil or gas, very few metal deposits of any consequence, and very little coal. Yet, the relatively modest (by Tokyo standards) hotel we stayed in was better than anything we had seen in the Soviet Union. The stores were filled with a great variety of quality merchandise, in stark and striking contrast to the situation in the USSR. And Tokyo's streets were filled with cars. What a difference, and why the difference? Two things caused this contrast: a relatively free, competitive economy, and an innovative population, very much oriented toward education. The Japanese have perhaps the most stringent and demanding educational system in the world. But it is still necessary for the educated Japanese, with their energetic economy, to have *something with which to work*; and, thus, access to minerals and energy minerals remains basic to survival.

Switzerland also is a country of very limited natural resources. Yet Switzerland enjoys one of the highest standards of living in the world. The source of this, again, is a free society, with a well educated population; but Switzerland also must be given access to mineral and energy mineral supplies. Free trade now allows this to happen.

One cannot make something out of nothing. The Swiss and the Japanese have to use steel to make their turbines. The Japanese could not be a world-class automobile manufacturer if it were not for imported iron ore. But perhaps if iron were not available, an innovative substitute could be found. The history of civilization has been marked by the discovery, development, and use (in a great variety of ways) of new resources to advance the human race. But what did it was the innovative and educated human mind.

The Indians of east Texas and the Eskimos of the Alaskan north slope walked over the two greatest oil fields in the United States. The Nomads of the Arabian Peninsula rode camels over the largest oil deposits in the world. But in each case, what could they have done about them, even if they had known they existed? Lacking knowledge that would unlock and utilize such resources, these people were poor. But the educated human mind changed all that. It is the combination of an educated human mind with some material resource, which makes the difference.

In more detail, the same situation applies. If crude oil is obtained from underground reservoirs or at natural oil seeps, as occurs many places in the world (almost all major oil fields exhibit oil seeps), it still has relatively few applications in its raw form. Its use has been somewhat erratic and site specific. For example, the natives, during Alexander the Great's Mideast campaign, poured crude oil down the road toward Alexander's headquarters, one night, and set it afire, to give the invader a hot reception.

But once technology is applied to this raw material, when the crude oil is put through a refinery and then some of these refined products are further processed, the end result is literally thousands of items which make for better living. The list is huge, including paint, a great variety of plastics, medicines, insecticides, special coatings of all kinds, inks, dyes, and many other things.

Similar situations exist among the metals, iron being an example. Occurring only very rarely in native form, iron was first found in meteorites, and swords were fashioned from this material, which were very highly prized for their potency in battle, and were termed "swords of heaven." Because of the very high melting point of iron, the metallurgy of iron was discovered and developed at a rather late date. But, finally, what were simply huge deposits of unusable iron ore, in many parts of the world, were then able to be exploited. The resource was there all the time but it took the human mind, with some degree of education, to make the iron available and useful.

Since the initial discovery that iron could be obtained from previously worthless rocks, it was further discovered that the addition of vanadium, chromium, tungsten, molybdenum, and other minor metals would give a variety of valuable properties to iron, producing specialty steels for many important and specialized purposes.

In Minnesota, on the Mesabi Iron Range, the rich hematite ore became exhausted, but very large quantities of lower grade ore, called taconite, exist. This low grade ore is now crushed and the iron separated out, concentrated into pellets and then shipped to steel mills. The uniform iron content of the pellets makes up, in part, for the initial lower grade ore, by allowing blast furnace operations to be more efficient, as compared with using raw but somewhat variable higher-grade ores. Although competition from foreign high-grade ores is severe, technology at least partially compensates for the depletion of the high grade ores of Minnesota and allows this area to continue competing to some degree, although iron mining is substantially reduced from what it once was. It was the educated human mind which allowed at least limited survival of this Minnesota industry.

There may be a limit, however, to how much the educated human mind can compensate (by ingenuity) for the scarcity or depletion of raw materials. Only the future can tell. Alchemy, in effect, can only proceed so far. You have to start with something; you cannot make something out of nothing. But so far, the educated human mind has been doing very well with resources that are tending to be of lower and lower grade. But there are limits to this trend also, and alternatives will have to be found.

Geological and geophysical exploration, over the past century, have given us a good inventory of the Earth's mineral resources; but we are using them up at an exponential rate. Since 1900, the human race has consumed more minerals and energy mineral resources than were used during all previous history. This consumption trend cannot be sustained indefinitely. There will come a day when there is no more oil available in the quantities we're are now using it; and this will also be true of other conventional resources such as copper, lead, zinc, tungsten, cobalt, antimony, and many other vital materials. Admittedly, iron and aluminum are in huge supply (aluminum, for example, makes up eight percent of the Earth's crust); but these metals take large amounts of energy to be recovered from their ores. The other metals cited are also still in reasonably ample supply; but as the higher grade and more accessible deposits are used, the cost in energy, in recovering more of these metals, will greatly increase. In some cases also, the ore bodies do not grade out, but end abruptly, in which circumstance the supply at that site is gone.

242

So the fundamental question is: Will it be possible for the educated human mind to find adequate substitutes for these critical materials? And in the energy field, can fusion ever be accomplished commercially? If not, as we turn inward and back to the other energy sources, can these be drawn on economically long enough to give us time to advance the technology of solar energy *conversion* and *use* sufficiently, so that the inexhaustible energy of the Sun can permanently take up a significant part of the energy load?

Can the educated human mind continue to carry us through the more difficult mineral resource times which lie ahead? Certainly there is no alternative but to try. The results will probably be a mixed situation, but there is no other option open to us. And if we try, we will surely achieve some level of success. Therefore, while we still have the abundance of energy and mineral resources which now allow only a small portion of the population — the energy producers, the miners, and the farmers — to supply the rest of us with the basics of life, we must make use of this affluent age to increase our knowledge of the physical world as much as we can, in order to prepare for the time when we will have to be much more ingenious than we are now, if we are to obtain the needed resources. What we will be doing then to accomplish this is difficult to say; but it is certain that the need to do so will definitely be there, and it will be done by the educated human mind.

As our mineral wealth, inherited from the geological past, is being depleted, the only wealth we can accumulate (rather than deplete) for the future is human knowledge. Like the Greeks, who were prevailed upon to forego their annual tribute of silver from the mines at Laurium, in order to build the fleet which defeated the Persians at the Battle of Salamis, so the human race as a whole must apportion a part of the current wealth toward the development of "shiploads" of information, which we hope will defeat the mineral resource problems of the future. Signs of these problems are already with us; for example, the passing by the United States of self-sufficiency in oil.

Directing our educational resources toward these ends is vitally important. Currently there is, in the United States, a very badly distorted allocation of educational facilities, and a misplaced emphasis. Japan trains 1,000 engineers for every 100 lawyers. In the United States, the ratio is just reversed. Every year in the United States, three times as many lawyers are graduated as exist in all of Japan. Japan is producing; the United States is suing. The United States is the most litigious nation in the world. The insurance group, Lloyds of London, reports that only 12 percent of its business is in the United States, but 90 percent of its insurance claims are there. It is very doubtful, however, that all the lawyers in the United States will ever discover an adequate substitute for oil — or how to reach the goal of commercially economic fusion power.

Another great and relatively non-productive drain on the time, talents, and money of U.S. citizens is the tax system. This not only discourages savings but involves a huge amount of legal and court time. At present, about 300,000 of our brighter minds (and it does take a bright mind) are involved in advising Americans on tax problems. There are more than 80,000 tax lawyers and accountants interpreting the complex tax laws for the citizenry. American taxpayers annually spend more than 300 million hours filling out tax forms, yet 40 percent find that they have to go to tax specialists, to complete the job. There is nothing comparable to this expenditure of time and money on tax matters, anywhere else in the world.

Our human resources need to be put to more productive use, building basic information for the future, not wrestling with lawsuits and tax forms. The United States has had the "luxury" of doing this up to now, because it had all the mineral and energy mineral resources it needed, for the most part. However, in the future, attention will have to increasingly turn toward providing for basic natural resources, including minerals, particularly as the population increases against diminishing supplies. The United States can no longer squander effort and time in the least productive enterprises. We would hope that the human mind will eventually devise a system which will give us an almost completely renewable, recyclable resource economy; but it is highly doubtful that this will be developed in a courtroom or in a tax accountant's office.

So far, the world has won this race between resources and the demand thereon. The natural bounty we inherited from all the geological events of the past, that produced the minerals and energy minerals we use today, combined with trained people, has allowed the world, in general, and western nations in particular, to continually and markedly raise their standard of living. But, unfortunately, the race does not stop — it continues indefinitely.

As high grade and abundant resources are used up, and, particularly, as that most important and convenient energy source of today, oil, is depleted, world-size questions loom before us: Can the increased demand, or even the steady demand, be met by the use of ingenuity in some fashion? Can adequate substitutes be discovered or devised? In other words, can knowledge in the future increase as fast as the need for knowledge to accomplish these tasks increases? So far, our knowledge of how to obtain and use our mineral and energy mineral resources has increased beyond our actual *need.* This has resulted in a higher and higher standard of living, as the developed countries have had energy and mineral supplies to use beyond those required for survival. But the easy technologies have been discovered and put in use, and they were used on the most easily recovered and richest mineral resources. Now we must perfect more difficult technologies, to use on more difficult energy mineral and mineral resource recovery problems.

In the particular case of energy, we believe we now know what the complete possible energy spectrum is, from simple wood burning to fusion. We have rather rapidly gone from wood burning to fission; but the step from fission to fusion is a very large one — probably as great, and possibly even greater, in terms of technology, than the total road from wood to fission. There is even some doubt, despite the "cold fusion" furor of the late 1980s, that fusion will ever be commercially accomplished. Thus, do we then, if we cannot achieve fusion, see the end of major advances and discoveries in energy sources? If we must depend on the known energy sources, without once again being able to discover and draw on new ones, as we did in 1859, when the Petroleum Interval was brought forth, what can the human mind do to solve this problem? This is a far greater challenge than we have faced in the past. However, there is no other tool with which to work, than the educated human mind. This has to be the ultimate resource and the ultimate strength of nations — the educated human mind, combined with material resources and a society in which knowledge can be freely pursued.

The statesman, Winston Churchill, perhaps noting the demise of the once geographically far-flung British Empire, observed that "The frontiers of the Twentieth Century are the frontiers of the mind." And so it will be from now on and indefinitely into the future. Still, after all is said and done, we must have the use of minerals and energy each day. So minerals and energy from their various sources will be important indefinitely, also, for the human mind must have materials with which to work. It cannot conjure something out of thin air; and even in recycling, there is an inevitable degradation and loss of some substance and energy. Entrophy is all about us.

Thus energy mineral and mineral resources must be combined with the ultimate resource, the educated human mind, to provide the basics for human existence: that will be available on into the indefinite future, shaping, we hope, the destinies of individuals and nations for the better.

To give this process a reasonable chance for success, we must all live together in peace. *Humanity now has too many problems in common to be divided.* And, individually, let us be kind to one another. It is the only way.

Bibliography

ABBOTT, J. F., 1916, Japanese expansion and American policies: The Macmillan Company, New York, 267 p.

ABELSON, P. H., 1987, Energy futures: American Scientist, New Haven, Connecticut, v. 75, p. 584-593.

AGNEW, A.F., (ed.), 1983, International minerals, a national perspective: Westview Press, Inc., Boulder Colorado, for American Assoc. for Adv. of Science, Washington, D. C,. 164 p.

ANDERSON, R. O., 1984, Fundamentals of the petroleum industry: Univ. of Oklahoma Press, Norman, Oklahoma, 390 p.

ANONYMOUS, 1981, Soviets push for construction of strategic Yamal gas line; Oil and Gas Journal, November. 2, Tulsa, Oklahoma, p. 57-59.

————, 1987, Venezuela advances toward goal of playing a major aluminum role: Engineering & Mining Journal, September, New York, p. 20-21.

————, 1987, New data lift world oil reserves by 27%: Oil and Gas Journal, December 28, Tulsa, Oklahoma, p. 33-75.

————, 1989, Shortages of fuel to trim China's oil exports: Oil and Gas Journal, April 3, Tulsa, Oklahoma, p. 28-29.

————, 1989, Soviet oil, gas industry languishes: Oil and Gas Journal, May 15, Tulsa, Oklahoma, p. 12-14.

————, 1989, U. S. oil supply/demand gap to widen: Oil and Gas Journal, May 22, Tulsa, Oklahoma, p. 28-29.

————, 1989, Qatar's huge North field is biggest Middle East gas project: Oil and Gas Journal, May 22, Tulsa, Oklahoma, p. 50-51.

————,1989, U. S. oil resource pegged at as much as 247 billion bbl: Oil and Gas Journal, June 5, Tulsa, Oklahoma, p. 32.

————, 1989, CERI: Canada's oil import dependence to grow: Oil and Gas Journal, September 4, Tulsa, Oklahoma, p. 27.

————, 1989, USGS, MMS cut estimates of U. S. resources: Oil and Gas Journal, September 4, Tulsa, Oklahoma, p. 28-29.

ARABIAN AMERICAN OIL COMPANY, 1968, ARAMCO handbook. Oil and the Middle East: Arabian American Oil Company, Dhahran, Saudi Arabia.

AYRES, EUGENE and SCARLOTT, C. A., 1952, Energy sources — the wealth of the world: McGraw-Hill Book Company, Inc., New York, 344 p.

BAKELESS, JOHN, 1920, Economic causes of modern war: A study of the period 1879-1918: Moffat, Yard, New York, 265 p.

BAKER, ARTHUR, et al., 1972, Forecasts for the future—minerals: Nevada Bureau of Mines and Geology Bull. 82, Reno, Nevada, 222 p.

BANNERMAN, H. M., 1957, The search for mineral raw materials: Mining Engineering, v. 9, p. 1103-1108.

BARNETT, H. J., and MORSE, CHANDLER, 1963, Security and growth : The economics of natural resource availability: Johns Hopkins Univ. Press, Baltimore, Maryland, 238 p.

BARNEY, G. O., 1980, The global 2000 report to the President of the U.S.: Pergamon Press, New York, 360 p.

BASIUK, 1977, Technology, world politics, and American policy: Columbia Univ. Press, New York 409 p.

BATES, R. L., and JACKSON, J. A., 1982, Our modern stone age: William Kaufman, Los Altos, California, 132 p.

BECHT, J. E., and BELLZUNG, L. D., 1975, World resource management: Key to civilizations and social achievement: Prentice-Hall, Englewood Cliffs, New Jersey, 329 p.

BEGAWAN, B. S., 1989, Brunei's work ethic: What, me worry? The Economist, February 25, p. 66-67.

BIRD, K. J., and MAGOON, L. B. (eds.), 1987, Petroleum geology of the northern part of the Arctic National wildlife Refuge, northeastern Alaska: U. S. Geological Survey Bull. 1778, Washington, D. C., 329 p., 5 pls. (maps).

BISHOP, J. E., 1989, Physicists outline possible errors that led to claims of cold fusion: Wall Street Journal, May 3, 1989, p. B4.

BOISSONAS, J., and OMENETTO, P., (eds.), 1988, Mineral deposits within the European Community; Special Publication 6, Society of Geology Applied to Mineral Deposits, published by Springer-Verlag, New York, 558 p.

BROBST, D. A., and PRATT, W. P. (eds.)., 1973, United States mineral resources: U.S. Geological Survey Prof. Paper 820, Washington, DC, 722 p.

BROOKS, D. A., and ANDERSON, P. W., 1974, Mineral resources, economic growth, and world population: Science, July, p. 13-18.

BROWN, D.S., 1989, Exploration versus exploitation: strategic mineral battles: Address (for U.S. Bureau of Mines) May 17, 1989 at 27th International Affairs Symposium, Lewis and Clark College, Portland, Oregon.

BROWN, HARRISON, 1978, The human future revisited: The world predicament and possible solutions: W. H. Norton & Company, New York, 287 p.

CAMERON, E. N. (ed.), 1972, The mineral position of the United States 1975-2000: Pub. for Soc. of Econ. Geologists Foundation, Inc., by Univ. Wisconsin Press, Madison, Wisconsin, 159 p.

—————, 1986, At the crossroads. The mineral problems of the United States: John Wiley and Sons, New York, 320 p.

CANNON, H. L., and DAVIDSON, D. F. (eds.), 1967, Relation of geology and trace elements to nutrition: Geol. Soc. of America Special Paper 90, Washington, D.C. 68 p., 1 pl. (map).

CASTLE, E. N., and PRICE, K. A., (eds.), 1983, U. S. interests and global natural resources, energy, minerals, food: Resources for the Future, Inc., Washington, D.C., 147 p.

CENTRAL INTELLIGENCE AGENCY, 1985, USSR energy atlas: Central Intelligence Agency, Washington, D.C., 79 p.

CHANDLER, W. U., et al., 1988, Energy efficiency: A new agenda: The American Council for an energy-efficient economy, Washington, D.C., 76 p.

CHEVRON CORPORATION, 1987, World energy outlook: Chevron Corporation, San Francisco, 13 p.

CHOUCRI, NAZLI, 1982, Power and politics in world oil: Technology Review, October, MIT Press, Cambridge, Massachusetts, p. 24-36.

CIPOLLA, C. M., 1974, The economic history of world populations: Sixth Edition: Penguin Books, Middlesex, England, 154 p.

CLAWSON, MARION (ed.), 1964, Natural resources and international development: Resources for the Future, Inc., Johns Hopkins Univ. Press, Baltimore, Maryland, 462 p.

COATS, D. R., 1985, Geology and society: Chapman and Hall, New York, 406 p.

COLE, J. P., 1972, Geography of world affairs: Penguin Books, Middlesex, England, 478 p.

CONGRESSIONAL QUARTERLY, April, 1974, The Middle East. U.S. Policy, Israel, oil and the Arabs: Congressional Quarterly, Washington, D.C., 100 p.

CONNELLY, PHILLIP and PERLMAN, ROBERT, 1975, The politics of scarcity: Resource conflicts in international relations: Oxford Univ. Press, London, 162 p.

COOK, JAMES, 1988, New player in aluminum: Forbes, February 8, p. 110, 112.

CORNOT, S., and VALAIS, M., 1988, Worldwide natural-gas trade, consumption to continue growth: Oil and Gas Journal, Tulsa, Oklahoma, June 6, p. 33-38.

CRAWFORD, JOHN, and SABURO, OKITA, 1978, Raw materials and Pacific economic integration: Univ. British Columbia Press, Vancouver, 343 p.

CRAWFORD, MARK, 1988, The mixed blessing of inexpensive oil: Science, v. 242, p. 1242-1243.

CROCKETT, R. N., et al., 1987, International strategic minerals inventory summary report; cobalt: U.S. Geological Survey Circular 0930-F, Washington, D.C., 54 p.

DALY, H.E. (ed.), 1973, Toward a steady-state economy: W. H. Freeman and Company, San Francisco, 332 p.

DAVIES, OLIVER, 1935, Roman mines in Europe: Oxford Univ. Press, London, 291 p.

DERRY, D. R., 1980, A concise world atlas of geology and mineral deposits: John Wiley and Sons, New York, 110 p.

deWILDE, J. C., 1936, Raw materials in world politics: Foreign Policy Reports, v. 12, September 15 p. 162-176.

DICKSON, DAVID, 1988, Norway: Boosting R&D for a post-oil economy: Science, v. 240, p. 1140-1141.

DORR, ANN, 1987, Minerals — Foundations of society (2nd ed.): American Geological Institute, Alexandria, Virginia, 96 p.

DUNKERLEY, JOY, et al., 1981, Energy strategies for developing nations: Pub. for Resources for the Future, Inc., by Johns Hopkins Univ. Press, Baltimore, Maryland, 265 p.

EARNEY, F. C. F., 1980, Petroleum and hard minerals from the sea: John Wiley and Sons, New York, 244 p.

ECKES, A. E., JR., 1979, The United States and the global struggle for minerals: Univ. Texas Press, Austin, Texas, 353 p.

EMENY, BROOKS, 1934, The strategy of raw materials: A study of America in peace and war: Macmillan, New York, 302 p.

ENGLISH, T. S. (ed.), 1973, Ocean resources and public policy: Univ. Washington Press, Seattle, Washington, 184 p.

FARRELL, CHRISTOPHER, 1989, It looks as if OPEC may have the last laugh; Business Week, September 25, New York, p. 34.

FEIS, HERBERT, 1944, Petroleum and American foreign policy: Stanford Univ. Press, Stanford, California, 62 p.

—————, 1950, The road to Pearl Harbor: Princeton Univ. Press, Princeton, New Jersey, 356 p.

FISCHMAN, L. L., 1980, World mineral trends and U. S. supply problems: Resources for the Future, Washington, D.C., 535 p.

FISHER, J. L. and POTTER, N., 1964, World prospects for natural resources: Resources for the Future, Inc., Johns Hopkins Univ. Press, Baltimore, Maryland, 73 p.

FISHER, W. L., and GALLOWAY, W. E., 1983, Potential for additional oil recovery in Texas: Univ. Texas Bureau of Economic Geology Circular 83-2, Austin, Texas, 20 p.

FLAWN, P. T., 1966, Mineral resources. Geology, Engineering, Economics, Politics, Law: Rand McNally & Company, New York, 406 p.

FRANK, TENNY, 1927, An economic history of Rome: Johns Hopkins Univ. Press, Baltimore, Maryland, 519 p.

GALL, NORMAN, 1986, We are living off our capital: (Interview with Joseph P. Riva Jr.) Forbes, New York, September 22, p. 62, 64-66.

GARTHOFF, R. L., 1959, The Soviet image of future war: Public Affairs Press, Washington, D.C., 137 p.

GEVER, JOHN, et al., 1986, Beyond oil. The threat to food and fuel in the coming decades: Ballinger Publishing Company, Cambridge, Massachusetts, 304 p.

GERASIMOV, I. P., ARMAND, D. L., and YEFRON, K. M., 1971, Natural resources of the Soviet Union: Their use and renewal: W. H. Freeman and Company, San Francisco, 349 p.

GILBERT, F. A., 1957, Mineral nutrition and the balance of life: Univ. Oklahoma Press, Norman, Oklahoma, 350 p.

GILDER, GEORGE, 1981, Wealth and poverty: Basic Books, Inc., New York, 306 p.

GORDON, DAVID and PANGERFIELD, ROYDEN, 1947, The hidden weapon: The story of economic warfare: Harper, New York, 238 p.

GORDON, R. B., et al., 1987, Toward a new Iron Age? Quantitative modeling of resources exhaustion: Harvard Univ. Press, Cambridge, Massachusetts, 173 p.

GREEN, STEPHEN, 1989, Soviets running out of oil: Moscow involvement in Middle East politics an indicator of domestic oil production woes:

Oregonian, Portland, May 7, p. G1, G4, reprinted in part from The Christian Science Monitor, April 6.

GRIGGS, G. B., and GILCHRIST, J. A., 1977, The Earth and land use planning: Duxury Press, Belmont, California, 492 p.

GUILD, O. W., 1974, Mineral resources of the Caribbean region: U.S. Geological Survey open-file report 74-363, Washington, D.C., 16 p.

HABASHI, F., and BASSYOUNI, F. A., 1982, Mineral resources of the Arab countries: Chemecon Publishing Limited, London, 60 p.

HARTLEY, J. N., et al., 1974, World mineral and energy resources – some facts and assessments: Battelle Pacific Northwest, Monograph 6, Richland, Washington, 137 p. and appendices.

HATFIELD, M. O., 1990, Fueling the future: Northwest Energy News, Northwest Public Power Planning Council, Portland, Oregon, v. 9, n. 2, p. 3-4.

HAYES, E. T., 1979, Energy resources available to the United States, 1985-2000: Science, v. 203, p. 233-239.

HEYNS, RONE, 1985, South Africa 1985 official yearbook of the Republic of South Africa: Republic of South Africa, Pretoria, 1062 p.

HIRSCH, R. L., 1987, Impending United States energy crisis: Science, v. 235, p. 1467-1473.

HINDMARSH, A. E., 1936, The basis of Japanese foreign policy: Harvard Univ. Press, Cambridge, Massachusetts, 265 p.

HODEL, DONALD, 1988, The right energy policy for America: Remarks by Interior Secretary Donald Hodel before the National Ocean Industries Association, April 11, Washington, D.C.

HOLMES, H. N., 1942, Strategic materials and national strength: The Macmillan Company, New York, 106 p.

HOLZER, T. L. (ed.), 1986, Man-induced land subsidence: Geological Society of America, Boulder, Colorado, 231 p.

HOWELLS, WILLIAM, 1981, The good fortune of Nauru: Harvard Magazine, p. 40-48.

HUBBERT, M. K., 1962, Energy resources. A report to the Committtee on Natural Resources of the National Academy of Sciences — National Research Council, Publication 1000-D: National Academy of Sciences — National Research Council, Washington, D.C., 141 p.

HUDDLE, FRANK, 1976, The evolving national policy for materials: Science, v. 191, p. 654-659.

HUNKER, H. L. (ed.), 1964, Erich W. Zimmerman's introduction to world resources: Harper & Row Publishers, New York, 220 p.

IKE, NOBUTAKA (ed.), 1976, Japan's decision for war: Records of the 1941 policy conferences: Stanford Univ. Press, Stanford, California, 306 p.

ISSAWI, CHARLES and YEGANEH, MOHAMMED, 1962, The economics of Middle Eastern oil: Frederick A. Praeger, New York, 225 p.

IVANHOE, L. F., 1988, Future global oil supply: American Assoc. Petroleum Geol. Bull., Tulsa, Oklahoma, v. 72, no. 3, p. 384.

—————, 1988, Future crude oil supply and prices: Oil and Gas Journal, Tulsa, Oklahoma, July 25, p. 111-112.

JAMES, DANIEL (ed.), 1981, Strategic minerals: A resource crisis: Council on Economics and National Security, Washington, D.C., 108 p.

JENSEN, J. T., 1988, World reserves won't limit international gas trade growth: Oil and Gas Journal, June 6, p. 38-42.

JENSEN, M. L. and BATEMAN, A. M., 1981, Economic mineral deposits, revised 3rd edition: John Wiley and Sons, New York, 593 p.

JONES, ARTHUR, 1976, The decline of capital: Thomas Y. Crowell Company, New York, 202 p.

KASH, D. E., et al., 1973, Energy under the oceans: Univ. Oklahoma Press, Norman, oklahoma, 378 p.

KELLER, E. A., 1979, Environmental geology: Charles E. Merrill Publishing Company, Columbus, Ohio, 548 p.

KENNEDY, PAUL, 1988, The rise and fall of the great powers: Random House, New York, 677 p.

KING, A. H. and CAMERON, J. R., 1975, Materials and new dimensions of conflict, in New dynamics in national strategy: Crowell, New York, 293 p.

KRAUS, ALAN, 1989, U. S. wrestles with need for energy policy: Investor's Daily, Los Angeles, California, August 7 p. 1, 34.

LAMEY, C. A., 1966, Metallic and industrial mineral deposits: McGraw-Hill Book Company, New York, 567 p.

LANDESBERG, H. H., FISCHMAN, L. L., and FISHER, J. L., 1963, Resources in America's future: Patterns of requirements and availabilities 1960-2000: Johns Hopkins Univ. Press, Baltimore, Maryland, 1017 p.

—————, 1964, Natural resources for U. S. growth. A look ahead to the year 2000: Published for Resources for the Future, Inc., by Johns Hopkins Univ. Press, Baltimore, Maryland, 260 p.

LAPP, R. E., 1973, The logarithmic century: Prentice-Hall, Inc., Englewood Cliffs, New Jersey, 263 p.

LEITH, C. K., 1931, World minerals and world affairs: A factual study of minerals in their political and international relations: McGraw-Hill, New York, 213 p.

—————, 1940, Peace — its dependence on mineral resources: Speech given at national Convention of League of Women Voters, New York City.

—————, FURNESS, J. W., and LEWIS, CLEONA, 1943, World minerals and world peace: The Brookings Institution, Washington, D.C., 254 p.

LEON, G. deL., 1982, Energy forever: Power for today and tomorrow: Arco Publishing, Inc., New York, 154 p.

LIBRARY OF CONGRESS, CONGRESSIONAL RESEARCH SERVICE, 1975, The development and allocation of scarce world resources: Washington, D.C., 399 p.

LOVERING, T. S., 1943, Minerals in world affairs: Prentice-Hall, Inc., New York, 394 p.

LUFTI, ASHRAF, 1968, OPEC oil: The Middle East Research and Publishing Center, Beirut, Lebanon, 120 p.

MACKENZIE, RICHARD, 1989, Soviet motives for the invasion [of Afghanistan] are starting to lose their veil: Insight, Washington, D.C., June 5, p. 30-31.

MACZAK, ANTONI, and PARKER, W. N. (eds.), 1978, Natural resources in human history: Resources for the Future, Inc., Washington, D.C., 226 p.

MAJORAM, TONY, et al., 1981, Manganese nodules and marine technology: Resources Policy, March, p. 45-57.

MALENBAUM, WILFRED, 1978, World demand for raw materials in 1985 and 2000: McGraw-Hill, Inc., New York, 126 p.

MANN, C. C., 1969, Abu Dhabi: Birth of an oil sheikhdom: Khayats, Beirut, Lebanon, 141 p.

MASTERS, C. D., 1981, Assessment of conventinally recoverable petroleum resources of Persian Gulf basin and Zagros Fold Belt (Arabian-Iranian basin): U.S. Geological. Survey open-file report 81-986, Washington, D.C., 7 p.

—————, and PETERSON, J. A., 1981, Assessment of conventionally recoverable petroleum resources Volga-Urals basin, U.S.S.R.: U.S. Geological Survey open-file report 81-1027, Washington, D.C., 7 p.

—————, 1981, Assessment of conventionally recoverable petroleum resources of the West Siberian basin and the Kara Sea basin, U.S.S.R.: U.S. Geological Survey open-file report 81-1147, Washington, D.C., 7 p.

—————, 1985, World petroleum resources—a perspective: U.S. Geological Survey open-file report 85-248, Washington, D.C., 25 p. gas, natural bitumen, and shale oil: Proceedings 12th World Petroleum Conference, John Wiley and Sons, New York, v. 5, p. 3-27.

McDIVITT, J. F., and MANNERS, GERALD, 1974, Minerals and men: Pub. for Resources for the Future, Inc., by Johns Hopkins Univ. Press, Baltimore, Maryland, 175 p.

McGREGOR, B. A., and OFFIELD, T. W., (no date), The exclusive economic zone: An exciting new frontier: U. S. Geological Survey general information publication, Washington, D.C., 20 p.

McKELVEY, V. E., and WANG, F. H., 1970, World subsea mineral resources: U.S. Geological Survey Map I-672, Washington, D.C.

McLAREN, D. J., and SKINNER, BRIAN (eds.), 1987, Resources and world development: John Wiley and Sons, New York, 940 p.

MERKLEIN, H. A., and HARDY, W. C., 1977, Energy economics: Gulf Publishing Company, Houston, Texas, 230 p.

MEYER, R. F. et al., 1984, Preliminary estimate of world heavy crude and bitumen resources: *in* The future of heavy crude and tar sands: McGraw-Hill, New York, p. 97-158.

MEYERHOFF, A. A., 1976, Economic impact and geopolitical implications of giant petroleum fields: American Scientist, New Haven, Connecticut, v. 64, p. 536-541.

—————, and MORRIS, A. E. L., 1977, Central American petroleum centered mostly in Mexico: Oil and Gas Journal, October 17, Tulsa, Oklahoma, p. 104-109.

—————, 1981, Energy base of the communist-socialist countries: American Scientist, v. 69, p. 624-631.

—————, 1981, Oil and gas potential of Soviet Far East: Scientific Press Ltd., Beaconsfield, Bucks, England, 176 p.

—————, and MEYER, R. F., 1984, Geology of heavy crude oil and natural bitumen in the USSR, Mongolia, and China: American Assoc. Petroleum Geol. Research Conference on Exploration for Heavy Crude oil and Natural Bitemen, Santa Maria, California, 293 p.

MILES, R. E., JR., 1976, Awakening from the American dream. The social and political limits to growth: Universe Books, New York, 246 p.

MILWARD, A. S., JR., 1967, Could Sweden have stopped the Second World War?: Scandinavian Economic History Review, v. 15, p. 127-138.

MORGAN, J.D., 1960, U. S. strategic materials stockpile and national strategy: Mining Engineering, v. 12, p. 925-928.

MORGAN, M. G. (ed.), 1975, Energy and man: Technical and social aspects of energy: The Institute of Electrical and Electronics Engineers, Inc., New York, 519 p.

MORRISON, S. E., 1965, The Oxford history of the American people: Oxford Press, New York, 1153 p.

MURDOCH, W. W., 1980, The poverty of nations: Johns Hopkins Univ. Press, Baltimore, Maryland, 382 p.

MYERS, C. V., 1985, World rollover: Falcon Press, Spokane, Washington, 280 p.

NATIONAL ACADEMY OF SCIENCES, 1969, Resources and man: W. H. Freeman and Company, San Francisco, 259 p.

—————, 1972, The Earth and human affairs: Canfield Press, San Francisco, 142 p.

NATIONAL STRATEGY INFORMATION CENTER, INC., 1980, The resource war and the U. S. business community: The case for a Council on Economics and National Security: Council on Economics and National Security, Washington, D.C., 42 p., appendix.

NETCHERT, B. C., and LANDSBERG, H. H., 1961, The future supply of major metals: a reconnaissance: Resources for the Future, Inc., Washington, D.C., 65 p.

NIXON, R. M., 1980, The real war: Warner Books, Inc., New York, 341 p.

NUSSBAUM, BRUCE, 1983, The world after oil. The shifting axis of power and wealth: Simon and Schuster, New York, 319 p.

ODELL, P. R., and ROSING, K. E., 1980, The future of oil. World oil resources and use: Kagan Page, London/Nichols Publishing Company, New York, 224 p.

—————, 1983, Oil and world power: Penguin Books, New York, 287 p.

OKITA, SABURO, 1974, Natural resource dependence and Japanese foreign policy: Foreign Affairs, v. 52, p. 714-724

O'TOOLE, JAMES, et al., 1976, Energy and social change: The MIT Press, Cambridge, Massachusetts, 185 p.

PARK, C. F., JR., 1968, Affluence in jeopardy. Minerals and the political economy: Freeman, Cooper & Company, San Francisco, 368 p.

—————, 1975, Earthbound. Minerals, energy, and man's future: Freeman, Cooper & Company, San Francisco, 279 p.

PAXTON, JOHN, 1984, The statesman's yearbook: St. Martin's Press, New York, 1692 p.

PETERSON, J. A., 1983, Petroleum geology and resources of southeastern Mexico, northern Guatemala, and Belize: U. S. Geological Survey Circular 760, Washington, D.C., 44 p.

PETROLEUM PUBLISHING COMPANY, 1977, Petroleum/2000: oil and Gas Journal, v.75, no. 35, Tulsa, Oklahoma, 538 p.

POLAND, J. F., and IRELAND, R. L., 1988, Land subsidence in the Santa Clara Valley, California, as of 1982: U. S. Geological Survey Professional Paper 0497-F, Washington, D. C., p. Fl-F6l.

POOL, ROBERT, 1988, Solar cells turn 30: Science, v. 241, p. 900-901.

—————, 1989, Fusion followup: Confusion abounds: Science, v. 244, p. 27-29.

—————, 1989, Skepticism grows over cold fusion: Science, v. 244, p. 284-285.

—————, and HEPPENHEIMER, T. A., 1989, Electrochemists fail to heat up cold fusion: Science, v. 244, p. 647.

POSS, J. R., 1975, Stones of destiny: Keystones of civilization: Michigan Technological Univ., Houghton, Michigan, 253 p.

PRICE, K. A., (ed.), 1982, Regional conflict & national policy: Resources for the Future., Inc., Washington, D.C., 142 p.

QUINN-JUDGE, PAUL, 1989, Soviet strike settlement sets costly precedent: The Christian Science Monitor, Boston, Massachusetts, July 26, p. 1-2.

RAYMOND, ROBERT, 1986, Out of the fiery furnace. The impact of metals on the history of mankind: The Pennsylvania State Univ. Press, University Park, Pennsylvania, and London, 274 p.

REES, DAVID, 1970, Soviet strategic penetration of Africa: Conflict Studies 77, p. 1-19.

REPETTO, ROBERT (ed.), 1985, The global possible: Resources, development, and the new century: Yale University Press, New Haven, Connecticut, 538 p.

RICH, NORMAN, 1973-1974, Hitler's war aims (2 vols.): Norton, New York, 584 p.

RIDGEWAY, JAMES, 1980, Who owns the Earth: Collier Books of Macmillan Publishing Company, Inc. New York, 154 p.

RIVA, J. P., JR., 1983, Assessment of undiscovered conventionally recoverable petroleum resources of Indonesia: Congressional Research Service, Library of Congress, report to the Congress, Washington, D.C., 23 p.

—————, 1987, Enhanced oil recovery methods: Congressional Research Service, Library of Congress, report to Congress, Washington, D.C., 14 p.

—————, 1987, The world's conventional oil production capabilities projected into the future by country: Congressional Research Service, Library of Congress, report to the Congress, Washington, D.C., 23 p.

—————, 1987, Fossil fuels: in Encyclopaedia Britannica, v. 19, p. 588-604.

—————, 1988, Oil distribution and production potential:, Oil and Gas Journal, Tulsa, Oklahoma, January 18, p. 58-61.

—————, 1988, Domestic oil production under conditions of continued low drilling activity: Congressional Research Service, Library of Congress, report to the Congress, Washington D.C., 12 p.

—————, 1989, Brazilian petroleum status: Congressional Research Service, Library of Congress, report to Congress, Washington, D.C., 21 p.

ROCKARD, T. A., 1932, Man and metals: A history of mining in relation to the development of civilization (2 vols.): McGraw-Hill, New York 1068 p.

ROCKEFELLER, NELSON, 1977, Critical choices for Americans. Vital resources—reports on energy, food, and raw materials: D. C. Heath, Lexington, Massachusetts, 187 p.

ROSEN, S. M. (ed.), 1975, Economic power failure. The current American crisis: McGraw-Hill, New York, 297 p.

ROSTOW, W. W., 1980, The world economy, history and prospect: Univ. Texas Press, Austin, Texas, 833 p.

ROUSH, G. A., 1939, Strategic mineral supplies: McGraw-Hill, New York, 485 p.

SAWYER, H. L., 1983, Soviet perceptions of the oil factor in U. S. foreign policy: The Middle East-Gulf Region: Westview Press, Boulder, Colorado, 183 p.

SCHROEDER, H. A., 1973, The trace elements and man: The Devin-Adair Company, Old Greenwich, Connecticut, 171 p.

SCHURR, S. H. et al., 1979, Energy in America's future: Pub. for Resources for the Future, Inc., by Johns Hopkins Univ. Press, Baltimore, Maryland, 555 p.

SHABAD, THEODORE and MOTE, V. K., 1977, Gateway to Siberian resources (The BAM): John Wiley and Sons, New York, 189 p.

SHIMKIN, D. B., 1953, Minerals — a key to Soviet power: Harvard Univ. Press, Cambridge, Massachusetts, 452 p.

SINCLAIR, J. E. and PARKER, ROBERT, 1983, The strategic metals war: Arlington House/Publishers, New York, 185 p.

SKINNER, B.J. (ed.), 1980, Earth's energy and mineral resources: William Kaufmann, Inc., Los Altos, California, 196 p.

—————, 1986, Earth resources (third edition): Prentice- Hall, Inc., Englewood Cliffs, New Jersey, 184 p.

SMITH, G. 0., 1927, Raw materials and their effect upon International relations: Carnegie Endowment for International Peace, 69 p.

SMITH, J.W., 1984, We need oil from oil shale: Mineral & Energy Resources, Colorado School of Mines, Golden, Colorado, v, 27, no. 6, p. 1-9.

SMITH, V. K. (ed.), 1979, Scarcity and growth reconsidered: Pub. for Resources for the Future, Inc., by Johns Hopkins Univ. Press, Baltimore, Maryland, 298 p.

SNYDER, G. H., 1966, Stockpiling strategic materials: Politics and national defense: Chandler Publishing Company, San Francisco, 314 p.

SPALLHOLZ, J. E., MARTIN, J. L., and GANTHER, H. E., (eds.), 1981, Selenium in biology and medicine: AVI Publishing Company, Inc., Westport, Connecticut, 573 p.

SPARROW, BARBARA (Chairman), 1987, Oil : Scarcity or security? Eighth report of Standing Committee on Energy, Mines, and Resources, Canada House of Commons, Ottawa, 121 p.

STALEY, EUGENE, 1937, Raw materials in peace and war: Council on Foreign Relations, New York, 326 p.

STEINHART, C. E. and STEINHART, J. S., 1974, Energy. Sources, uses, and role in human affairs: Duxbury Press, Wadsworth Publishing Company, Inc., Belmont, California, 362 p.

STOBAUGH, ROBERT, and YERGIN, DANIEL (eds.), 1983, Energy future (3rd ed.): Random House, New York, 459 p.

STOESSINGER, J. G., 1969, The might of nations: Random House, New York, 461 p.

—————, 1974, Why nations go to war: St. Martin's Press, New York, 230 p.

STRAUSS, S. D., 1983, The quest for gold and silver: Including a history of the interaction of metals and currency: Mineral & Energy Resources, Colorado School of Mines, Golden, Colorado, v. 26, no. 6, p.1-10.

SURO, ROBERT, 1990, 3 states, once fat on oil, try to break with past: New York Times, New York, January 14, p. 1, 14.

SUTLOV, ALEXANDER, 1972, Minerals in world affairs: Univ. Utah Printing Services, Salt Lake City, Utah, 200 p.

SYMONDS, W. C., et al., 1988, Kuwait's money machine comes out buying: Business Week, New York, March 7, p. 94-95, 98.

TANZER, MICHAEL, 1980, The race for resources. Continuing struggles over minerals and fuels: Monthly Review Press, New York/London, 285 p,

TARABRIM, E. A., 1974, The new scramble for Africa: Progress Publishers, Moscow, USSR, 234 p.

TEMPLE, JOHN, 1972, Mining. An international history: Praeger Publishers, New York, 144 p.

THEOBALD, P. K. et al., 1972, Energy resources of the United States: U. S. Geological Survey Circular 650, Washington, D.C., 27 p.

UNDERWOOD, E. J., 1962, Trace elements in human and animal nutrition: Academic Press, Inc., New York, 429 p.

UNITED STATES BUREAU OF MINES, 1983, The domestic supply of critical minerals: Washington, D.C., 49 p.

————, 1985, Mineral facts and problems: U. S. Bureau of Mines Bull. 675, Washington, D.C., 956 p.

————, 1989, Mineral commodity summaries 1989: U. S. Bureau of Mines, Washington, D.C., 191 p.

VAN DER MERWE, N. J., and AVERY, D. H., 1982, Pathways to steel: American Scientist, v. 70, p. 146-156.

VOSKUIL, W. H., 1955, Minerals in world industry: McGraw-Hill, New York, 324 p.

WALLACE, H. A., 1941, The silent war: Battle for strategic materials: Colliers, v. 108 (November 22), p. 14-15.

WALLACE, MACK (chairman), 1987, Texas Railroad Commission, Oil and Gas Division, 1986 Annual Report to the Governor: Austin, Texas.

WARD, J. M., et al., 1975, In short supply. A critical analysis of world resources: National Textbook Company, Skokie, Illinois, 342 p.

WARKEN, H. V., 1965, Medical geology and geography: Science, v. 148, p. 534-536.

WARREN, KENNETH, 1973, Mineral resources: John Wiley and Sons, New York, 272 p.

WATKINS, T. H., 1971, Gold and silver in the West: American West Publishing Company, Palo Alto, California, 287 p.

WELLS, H. G., 1920, The outline of history: Garden City Publishing Company, Inc., New York, 1171 p.

WHITMORE, F. C., JR., 1981, Resources for the 21st century: Summary & conclusions of the International Centennial Symposium of the U.S. Geological Survey: U.S. Geological Survey Circular 857, Washington, D.C., 41 p.

WILLIAMS, BOB, 1989, Alaskan oil flow outlook dimmed by tax threat: Oil and Gas Journal, Tulsa, Oklahoma, October 16, p. 26, 28.

WILLRICH, MASON, 1975, Energy and world politics: The Free Press, Macmillan Publishing Company, Inc., New York, 234 p.

WILSON, C. L. (project director), 1977, Energy: Global Prospects 1985-2000: McGraw-Hill, New York, 291 p.

WOLFE, J. A., 1984, Mineral resources, a world review: Chapman and Hall, New York, 293 p.

WORLD RESOURCES INSTITUTE, and INTERNATIONAL INSTITUTE FOR ENVIRONMENT AND DEVELOPMENT, 1986, World resources: Basic Books, Inc., New York, 353 p.

WU, YUAN-LI, 1952, Economic warfare: Prentice-Hall, Inc., New York, 403 p

YULSMAN, TOM, 1988, Light at the end of the tunnel: Discover, New York, November, p. 90-92, 94-95.

YOUNGQUIST, WALTER, 1980, Investing in natural resources (2nd ed.): Dow Jones-Irwin, Inc., Homewood, Illinois, 281 p.

Index

A

C

D

J

K

N

O

P

S

T

U